OXFORD MEDICAL PUI

ALLERGY

THE FACTS

ALSO PUBLISHED BY OXFORD UNIVERSITY PRESS

Ageing: the facts
Nicholas Coni, William Davison,
and Stephen Webster

AIDS: the facts
A. J. Pinching (forthcoming)

Alcoholism: the facts
Donald W. Goodwin

**Arthritis and rheumatism:
the facts**
J. T. Scott

Asthma: the facts
Donald J. Lane and Anthony Storr

Back pain: the facts
Malcolm Jayson

**Blindness and visual handicap:
the facts**
John H. Dobree and Eric Boulter

Blood disorders: the facts
Sheila T. Callender

Breast cancer: the facts
Michael Baum

Cancer: the facts
Sir Ronald Bodley Scott

Childhood diabetes: the facts
J. O. Craig

Contraception: the facts
P. Bromwich and A. D. Parsons

**Coronary heart disease: the
facts**
J. P. Shillingford

Cystic fibrosis: the facts
Ann Harris and Maurice Super

Eating disorders: the facts
(second edition)
S. Abraham and D. Llewellyn-
Jones

Epilepsy: the facts
Anthony Hopkins

Hip replacement: the facts
Kevin Hardinge

Hypothermia: the facts
K. J. Collins

Kidney disease: the facts
(second edition) Stewart Cameron

**Liver disease and gallstones:
the facts**
A. G. Johnson and D. Triger

Lung cancer: the facts
C. J. Williams

Migraine: the facts
F. Clifford Rose and M. Gawel

Miscarriage: the facts
G. C. L. Lachelin

Multiple sclerosis: the facts
(second edition) Bryan Matthews

Parkinson's disease: the facts
Gerald Stern and Andrew Lees

Phobia: the facts
Donald W. Goodwin

Rabies: the facts
(second edition) Colin Kaplan,
G. S. Turner, and D. A. Warrell

Schizophrenia: the facts
Ming Tsuang

**Sexually transmitted diseases:
the facts**
David Barlow

Stroke: the facts
F. Clifford Rose and R. Capildeo

Thyroid disease: the facts
R. I. S. Bayliss

ALLERGY

THE FACTS

ROBERT DAVIES and SUSAN OLLIER

*St Bartholomew's Hospital,
London*

OXFORD   NEW YORK   TOKYO
OXFORD UNIVERSITY PRESS
1989

Oxford University Press, Walton Street, Oxford OX2 6DP
Oxford New York Toronto
Delhi Bombay Calcutta Madras Karachi
Petaling Jaya Singapore Hong Kong Tokyo
Nairobi Dar es Salaam Cape Town
Melbourne Auckland
and associated companies in
Berlin Ibadan

Oxford is a trade mark of Oxford University Press

Published in the United States
by Oxford University Press, New York

British Library Cataloguing in Publication Data
Davies, Robert J.
Allergy
1. Man. Allergies
I. Title II. Ollier, Susan
616.97
ISBN 0–19–261439–8
ISBN 0–19–261858–X (pbk)

Library of Congress Cataloging in Publication Data
(Data available)

Set by Latimer Trend & Company Limited, Plymouth

Printed in Great Britain by
Biddles Ltd
Guildford and King's Lynn

PREFACE

Everyone is allergic to something, even if it is only Monday mornings, and it is difficult to get through a pleasant meal with friends without a mention of allergy to food or drink. Supermarkets list the additives and preservatives present in or absent from their produce, and some people feel better if the food they eat is organically grown and purchased from a health shop. Allergy to wines causes headaches and allergy to shellfish skin rashes. Cosmetics are described as hypo-allergenic and duvets and pillows as non-allergenic. Allergy to the humble bee can kill, and allergy to medicines makes the cure worse than the illness. One in seven primary school children carries an inhaler to alleviate wheezing, and for many of us the all too few hot summer days are clouded by sneezing and a blocked or runny nose.

There is no doubt that allergy is a very common problem and that it is affecting more and more people. Surprisingly, in Britain allergy has aroused relatively little interest amongst the medical profession, and in consequence sufferers have sought help from practitioners of alternative and fringe medicine. Articles on allergy abound in popular magazines and newspapers. Talking, as we often do, to non-medical audiences it is clear that while interest in the subject is intense, knowledge is scant and it is never possible to answer all the questions raised and deal with all the misconceptions in the time available. The only answer was to write this book. Allergy is complex but can be understood. For those who want to know more of the underlying science, details can be found in the Appendices. Our aim is to help you understand what you can be allergic to and what you can't, what diseases are caused by allergy, which symptoms can be alleviated and which can be cured. If we can persuade you that you are unlikely to be allergic to the twentieth century we will be satisfied, and so, we hope, will you.

St Bartholomew's Hospital   R.D.
December, 1988   S.O.

ACKNOWLEDGEMENTS

We are most grateful to our colleagues in the Department of Respiratory Medicine of St Bartholomew's Hospital for their contributions: to Diana Cundell, Research Assistant, for preparing the artwork; to Lucille Daniels, State Registered Dietician, for the section on exclusion diets; to Karen Henley, Research Assistant, for the section on allergen avoidance; and to Laura Roberts, Research Secretary, for typing the manuscript.

CONTENTS

Part I

ALLERGY AND ALLERGENS

1

THE NATURE OF ALLERGY

We all know what we mean by allergy, or do we? It is really a question of definition. The word is derived from the Greek and means simply an altered reaction, a definition which itself opens the door to the widest interpretation. Popularly, allergy is thought of as a curious over-reaction to substances outside our bodies and, in extreme cases, to everything in the external world—hence the concept of allergy to the twentieth century, the only cure for which is to live in an isolated room breathing filtered air and eating organically grown foods free of any colourings or preservatives. Nowadays people often say that they are allergic to things which make them feel unwell, particularly if no other cause can be found for their symptoms. Even doctors resort to the phrase 'it must be an allergy' when no other diagnosis can be made. The facts are that people can be allergic to the most bizarre substances and that allergy is extremely common. It also hits the headlines, as in the case of one of our patients, who suffers from an extremely severe allergy to potatoes and featured recently in the *News of the World* under the caption: 'Half a spud and she'd have had her chips'. With all the current interest in allergy to foods and additives, it is important to get a proper perspective. The commonest allergies are those caused by what we inhale, the best-known examples being hay fever and asthma, diseases which are definitely on the increase. One of the difficulties in understanding allergy, and indeed diagnosing it, is that there is often a grain of truth in the wildest claims. People often joke that they are allergic to work or Monday mornings, but sometimes they really are, as is explained in Chapter 6.

The effect of allergies varies from mildly disabling to fatal. Fortunately the majority only debilitate, for example hay fever, rather than kill, as in the relatively rare cases of allergy to bee or wasp stings. Debility, however, is relative. Hay fever can be extremely disruptive when it strikes students sitting summer-time exams, and asthma can rob children of the pleasures of physical

HALF A SPUD AND SHE'D HAVE HAD HER CHIPS!

KILLER POTATO

Joyce Evans is a cook – but cooking puts her life in danger. If she so much as smelt or touched the potatoes she has to serve up every day, she'd fall seriously ill – and eating half a spud would kill her. Her weird allergy means that she never dares eat in a restaurant. She can't even visit friends without warning them well in advance.

Joyce, 50, suffers from a rare and very violent reaction to potatoes – just touching a raw potato brings her out in blisters and steam from cooking potatoes makes her lips and face swell and her throat close up, choking her. "It's like having a tennis ball in your throat and fighting for breath," says Joyce, who works as cook and housekeeper in an office block near London's Fleet Street where she has to make lunch for up to 20 people every day.

The first time Joyce realised something was wrong was when she was on holiday in Spain. She ate some mashed potato – the first time she'd had potato for some time because she'd been on a diet – and her throat immediately started to blister. That time she got better by herself but, back at home again, she noticed that it was getting harder and harder to cook potatoes. If she got any of the water on her face, her eyes started to blister and her chest became very tight "just like asthma".

Then one evening she ate a roast beef dinner – and the bit of potato she ate provoked a terrible reaction. Less than half an hour after the meal her entire body and face were covered in huge hard lumps. She called her doctor who gave her a big dose of antihistamines to bring the swellings down.

Smaller and smaller quantities of potato were now causing ever more serious attacks, so Joyce made the decision to give them up completely. But that was when she suffered her worst attack ever, so bad she thought she was going to die – brought on by a piece of chocolate. She broke off a piece from a bar her husband, Philip, had left unfinished on the sideboard and popped it into her mouth. But

> ### 'If I touch a raw potato, I blister, my face swells and I choke as my throat closes up'

Philip, who's the block's caretaker, had been eating crisps – and a tiny crisp crumb stuck to the chocolate made her tongue and lips swell up terrifyingly.

The attack lasted 48 hours. "I was so frightened I was going to choke I sat up all night and the next day my GP told me to go to hospital," Joyce recalls. In hospital the doctors put her through a series of tests to rule out the possibility that her allergy might all be in the mind. They put her behind a screen, with a clip on her nose, in case the association of certain smells set her off, and made her suck a peppermint while consultant physician Dr Robert Davies and his colleagues cooked a variety of vegetables on the other side of the screen, a few feet away.

"As soon as we started cooking the potatoes, there was a quick response," Dr Davies recalls. Her lungs went down to two-thirds of their usual power. They blindfolded Joyce and peeled a potato in front of her – and got exactly the same reaction. But they found that no other food provoked any kind of allergic reaction in her.

Potato allergy is extremely rare, "about one in a million", says Dr Davies. "Joyce is the first case I've seen in 20 years' practice. And the severity of her reaction makes her even more unusual. Half a potato would probably kill her," he warns. But there is nothing the doctors can do for her. All Dr Davies could recommend was that Joyce should avoid potatoes completely.

And that is just what Joyce and Philip have done. Philip, 67, has come to the rescue. He's fitted out a small hut on the flat roof outside their flat with a camping gas stove and that's where he goes to cook all of the potatoes Joyce needs for the lunches she does each day.

Once cooked, drained and left in a covered dish for at least five minutes, the potatoes are then "safe" for Joyce to take to work. Philip used to cook them in their own flat with the kitchen door shut. But, as her condition got worse, she became more and more sensitive to the steam seeping round the door.

Now that she can't even go into a house where potatoes are being boiled, some of her relatives have taken to doing all their cooking in the garages when they know she's coming round.

Joyce's allergy means that they've given up eating out. And, on holiday, they go camping. "Restaurants are just too much trouble – you can never be sure there isn't potato in a soup or sauce. And it's difficult to make the staff understand the severity of the problem... If I'm served vegetables with a spoon that's just touched potatoes, my mouth immediately starts to blister," she says.

Joyce thinks she's come to terms with her strange problem, though not without regrets. "Rice isn't much of a substitute for the potatoes I enjoyed without problems all those years," she sighs wistfully. "But I make do and I'm delighted I didn't have to give up my job."

Report by Joanna Lyall

> ● **Rooftop cookery...**
> "I do lunch for about 20 every day but, because I can't go near potatoes, it's my husband who has to cook them – in a shed on the roof, outside our flat. Once cooked and left in a covered dish for five minutes, they're safe – just so long as I don't actually come into contact with them..."

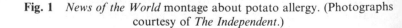

27

Fig. 1 *News of the World* montage about potato allergy. (Photographs courtesy of *The Independent*.)

activity. Fortunately, public interest in allergy has coincided with an explosion of knowledge about what causes it, how it develops, and what can be done to treat it. Allergy results from extremely complex mechanisms, and how many of the treatments work remains unknown. The hope is that the more becomes known about allergy the more rational and effective therapy will become.

A growing problem?

One in five teenagers suffers from hay fever and one in seven primary school children has asthma. There are over half a million asthma sufferers of all ages in Britain and hay fever is at least twice as common. No one quite knows how frequent allergic skin diseases and food allergy are, but eczema in infants (often related to cow's-milk allergy) is a frequent problem. These facts underline the importance of allergies to the nation's health, and yet the problem is greater than this since there is growing evidence that asthma and allergies are getting commoner. Extensive studies in general practice have shown that the number of people attending their doctors for treatment for both asthma and hay fever has doubled in the last 10 years. Allergies are particularly common in certain parts of the world. In China and Hong Kong, for example, asthma is only one-twentieth as common as in the United Kingdom, but in New Zealand and Australia asthma occurs five times as frequently as in Britain. The reasons for these differences are complex, but are probably influenced by race and the economic development of the country.

The importance of heredity in the development of asthma is clearly shown amongst the inhabitants of Tristan da Cunha. Asthma was common amongst the early settlers and now half the island's population suffer from asthma. Similarly, the high incidence of asthma amongst the Maoris and their subsequent inter-marriage with European immigrants may have helped to account for the higher frequency of asthma amongst New Zealanders. Studies on the inhabitants of New Guinea have shown the importance of lifestyle in the development of allergies and asthma, in that a change to more Westernized ways of living has been associated with an increase in allergic diseases.

Asthma may be even more common than is thought. Recent studies have suggested that as many as a quarter of the population may be hidden asthma sufferers, with their symptoms, for example prolonged coughing after chest colds, remaining unrecognized as being due to asthma. In addition, almost half the adult population of Britain react to common allergens, at least on skin-testing (see Chapter 9).

Why is allergy an increasing problem?

Speculation is always dangerous but it may be that the increase in allergy is in some way related to our healthier lives. The basis of allergy lies in the immune system which originally developed to protect us from invading viruses, bacteria, and parasites. It is possible that with the decrease in importance of infectious diseases in the developed world, the immune system is reacting increasingly to other foreign substances, for example pollens, which are themselves inately harmless but which are now such important causes of allergic diseases.

A matter of definition

The first person to use the term 'allergy' was a doctor working in Vienna at the turn of this century, when dramatic advances were being made in vaccination. Today we are very familiar with the benefits of this type of treatment against both bacterial and viral diseases, for example immunization against whooping cough, German measles, polio, and smallpox. The injection of dead or modified bacteria and viruses leads to the development of immunity, i.e. protection against further infections. What was realized in the early twentieth century was that too many injections caused the individual to become over-immunized and to react against the injections themselves. This over-reactivity was called allergy. How immunity and allergy develop in the human body is extremely complicated. Those readers who are interested in the mechanisms of immunity and how they were discovered, will find details in Appendix A. Suffice it to say that allergy is a form of immunity in which the natural defence mechanisms of the body have misfired, and instead of protecting actually cause disease.

**Allergy is the inappropriate and harmful response
of the immune system to normally harmless substances**

Grass pollen is quite harmless but allergy to it can cause disabling

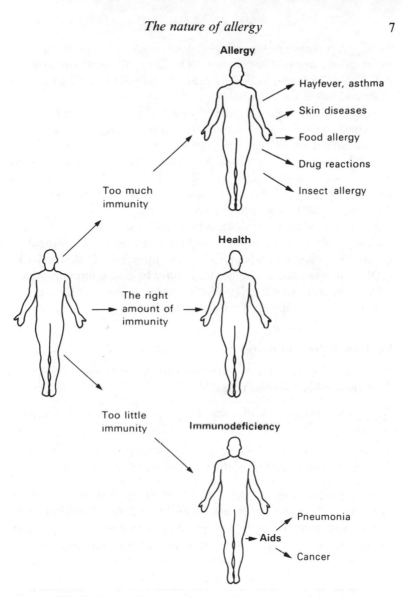

Fig. 2   Immunity: too much; too little; right amount.

nasal symptoms and wheezing and shortness of breath which at worst causes death. The way in which allergy to the harmless grass pollen develops and how this leads to hay fever and asthma is described in the next chapter.

Today we are faced with the prospect of a new plague sweeping the world caused by the human immuno-deficiency virus (HIV). In the acquired immuno-deficiency syndrome (AIDS) the body exhibits reduced rather than heightened reactivity to the external world—the exact opposite to allergy. When AIDS is fully developed, organisms which are normally harmless are able to grow inside the body with disastrous results. One such microscopic organism, which exists as a tiny cyst in dust that we inhale, is the cause of death in 85 per cent of AIDS patients. Normally we readily exterminate this organism, called *Pneumocystis cariniae*, but when immunity is destroyed by HIV, fatal pneumonia results. A correctly balanced immune system is vital for a healthy life. Too little immunity, as in AIDS, causes death, and too much immunity causes allergy.

Cardinal features of allergy

(1) Allergy only develops after the substance causing it has been encountered several times;

(2) only a proportion of those exposed to the substance develop disease;

(3) it involves special proteins (antibodies) present in the body (these are described both in Appendix A and in Chapter 2);

(4) powerful chemicals are released from specialized cells (mast cells) in the body (the nature of these chemicals and the cells releasing them are fully described in Chapter 2).

To the lay person 'allergy' is usually used in a much less precise way and includes:

1. Allergy to work. What we really mean is that we have a psychological aversion to going to work (although occasionally some people really are allergic to chemicals found in the work environment).

2. Allergic to coffee. What we really mean is that we get excited, can't sleep, and are restless due to the effects of the drug caffeine.

3. Allergy to food additives. Even this is not true allergy since it does not involve antibodies and mast cells (see Chapter 2).

4. Allergy to milk. This covers a multitude of different reactions. Sometimes allergy is involved, but often there is a lack of an enzyme which breaks down milk so that it can be absorbed. If the enzyme is missing, as in most Chinese, milk isn't absorbed, causing diarrhoea.

5. Allergy to the weather. This is very complex. Virtually everyone feels lethargic and out of sorts on hot and humid days. However, some drugs which cause skin diseases are activated by sunshine; and other skin rashes, for example urticaria, are brought on when the body heats up. Cold, dry air causes most asthmatics to develop an attack of wheezing, not as a result of an allergic reaction but due to cooling and drying of the lining of the lungs.

6. Allergy to cigarette smoke, aerosols, strong perfumes, and disinfectants. Again, sufferers from hay fever and asthma experience worsening symptoms when they encounter these substances. This is not an allergic reaction but the result of irritation of the more sensitive lining of the nose and lungs of allergy sufferers.

7. Allergy to shellfish. Even this isn't an allergy in the true sense of the word. Many shellfish contain substances which cause mast cells to liberate their powerful chemicals without involving antibodies.

Sorting out allergy is far from easy and full of pitfalls. Read on and you will understand more.

What causes allergy?

The substances in the environment which cause allergy are usually proteins, and are called allergens. The number of potential allergens is almost limitless but we now know quite a lot about the most important allergens causing disease in man. They are:

● mites
● pollen
● moulds
● domesticated animals
● insects

- industrial chemicals
- medicines
- foods

These are fully described in Chapter 3.

The most important allergen world-wide is a protein called 'Der p1' found in the droppings (faeces) of the house dust mite, which is found in enormous numbers. This tiny translucent creature, 0.35 mm in size, lives in damp houses and eats dead skin scales. It is found in maximum numbers in bedding. Apart from the house dust mite, there are similar mites that live on vegetable material and which can be found in grain stores, and there are even mites that live by preying on other mites. If a chemical could be found which would eradicate mites while causing no harm to insects, animals, and man, the frequency of allergic diseases would drop dramatically.

Pollens abound all over the world but the type differs from country to country. In the United Kingdom, timothy and rye grass are the commonest species, whereas in Scandinavia tree pollens, particularly birch, are the most important cause of allergic rhinitis. Although grasses cause hay fever in the summer months in America, the major problem is allergy to ragweed, which causes very intense symptoms in the early autumn.

Nose and chest diseases are particularly common in Northern Europe where the damp weather encourages the growth of mites and also of moulds. Penicillin moulds abound in damp houses, but appear to be a rare cause of allergy, whereas other moulds such as *Aspergillus fumigatus*, *Cladosporium*, and *Didymella* can cause disabling lung and nose disease. Pets are considered an important part of the British home and yet cause considerable misery, particularly in children, as they can be the cause of severe asthma and eczema.

In recent years it has been appreciated that there are over 200 industrial dusts, vapours, or fumes which can cause asthma at the workplace. Careful study of individuals who have contracted asthma as a result of exposure to these chemicals has shown that once asthma has developed it can progress, even though the individual is no longer exposed to the causative allergen.

The fact that medicines, particularly antibiotics, can cause allergy is well known, but it must be remembered that other drugs can have a similar effect. Those allergic to medicines should wear a Medic Alert bracelet or pendant (Medic Alert Foundation, 11–13 Clifton

Terrace, London N4. Tel: 01-236-8596) to warn doctors in case of an accident.

Facts and fantasies

Allergies are often difficult to diagnose, and it is equally difficult to prove that a person's symptoms are not due to allergy. Allergy, particularly to foods, has been suggested as the cause of rheumatoid arthritis, ulcerative colitis, Crohn's disease, multiple sclerosis, and hyperactivity in children. At present there is no proof that food allergy causes any of these conditions. Nevertheless, it is understandable that many individuals suffering from chronic diseases think that a change in their diet could be the cure they have been waiting for. Unfortunately, this has resulted in numerous clinics and laboratories being set up claiming to diagnose and offer treatment for food allergy. It can cost anything up to £100 for tests on blood or locks of hair and the results are unreliable. An investigation by *WHICH?* magazine found that private allergy clinics gave different results for tests on the same person, and concluded that these clinics did not reliably identify allergies.

What can be done to treat allergies?

The most important aspect of the treatment of allergic diseases is to confirm that allergy is involved and to determine the nature of the allergens. Where the allergen is the domestic pet or a material at work early recognition and avoidance can alleviate symptoms and lead to a cure, but all too often symptoms are the result of allergy to many different allergens and avoidance alone is rarely sufficient. The practice of desensitization with allergy extracts started in the United Kingdom, but has become less used in this country in the past 20 years with the development of effective and safe drugs for the management of allergic diseases. It has been claimed, though never proved, that desensitization injections can cure allergic diseases. However, modern drugs can alleviate symptoms dramatically, although they do not lead to a permanent cure. These drugs and desensitization injections, and their pros and cons, are fully described in Chapter 11. Although great strides have been made in our understanding of the mechanisms leading to allergy, more needs to be known before we are able to fine-tune the body's immune response to maintain protection but avoid allergy.

2

ALLERGY, INFLAMMATION, AND THE IMMUNE RESPONSE

A little history

Understanding the mechanisms involved in allergy is not easy; what we know today has gradually evolved over the past 100 years, although it is fair to say that there is still much more to learn. The modern foundations of allergy began with the brilliant deductions of a Manchester physician called Charles Harrison Blackley towards the end of the nineteenth century. As is so often the case, Charles Blackley's interest in this subject stemmed from the fact that he himself suffered from hay fever and asthma and was able to use himself as his own 'guinea-pig'. He noticed that he would start sneezing and wheezing when he went into his study where fresh grasses had been placed in a vase. His first experiment was simple. He shook pollen from the grasses and placed some into his eye which immediately began to water. Next he tried inhaling the pollen grains. This made him wheeze, showing quite conclusively that pollen grains were the cause of his summer-time sneezing and wheezing. Next he showed that a watery extract of pollen grains scratched into the skin caused redness, itching, and wealing. This was the beginning of allergy skin-testing, which is fully described in Chapter 9. It was soon discovered that this type of skin reaction only occurred in those people who suffered from hay fever, and the next question that needed to be answered was why hay-fever sufferers reacted in this way.

The next milestone in our understanding came from another doctor, Heinz Kustner, who also suffered from allergy. He was allergic to fish, and would develop swelling of the lips and tongue on eating it. He and his colleague, Carl Prausnitz, working in the 1920s, set out to investigate this further. An extract of fish injected into Dr Kustner's skin gave rise to the characteristic redness and blistering. This did not happen if fish extract was injected into the skin of Dr

Prausnitz, who was not allergic to fish. They deduced that there was something different in Dr Kustner's blood which might account for his allergy. They took some of his blood, separated out the cells, and injected the remaining serum into Dr Prausnitz's skin. Lo and behold, when fish extract was subsequently injected into the site where Dr Kustner's serum had previously been placed the redness and blistering occurred. This was a remarkable discovery, although it took a further 40 years before the 'something' present in the blood of allergic people was finally identified.

In those intervening years much had been learnt about the immune system which protects us against bacteria, viruses, and parasites. The major proteins present in blood and tissue fluid, called antibodies, which protect us against invading organisms, had been identified and fully characterized. They were found to belong to three major groups, called immunoglobulin G (IgG), immunoglobulin M (IgM), and immunoglobulin A (IgA) (see Appendix A for a full description). None of these antibodies, however, could transfer allergy to a non-allergic individual in the way described by Prausnitz and Kustner.

It was the Japanese who made the next great contribution to our understanding of allergy. Just over 20 years ago Dr Ishizaka, working in America, found tiny traces of a fourth group of immunoglobulin in blood which could transfer allergy. This group of antibodies is now called immunoglobulin E (IgE). The structure of the antibody is shown in Fig. 39. Basically it is just like other immunoglobulins in that one end of the protein molecule will combine quite specifically and exclusively with one type of protein present on the allergen with which it comes into contact. IgE was, however, found to be slightly different from the other immunoglobulin groups in that it had an extra piece on the opposite end which made it stick firmly to two closely related cells, one found in blood (called a basophil), and the other in the surface linings of the body (called a mast cell). These cells were first discovered in the late nineteenth century by a German doctor, Paul Ehrlich. He described the major characteristics of these cells, i.e. that they are packed with granules, as shown in Fig. 3. We now know that these granules contain powerful chemicals—histamine in particular.

Histamine was first discovered in 1910 by an Oxford scientist, Henry Dale. He showed that this substance caused powerful contraction of certain muscles in the body but relaxed others. It had no effect on the sort of muscle we use to move our limbs, but did contract the type of muscle present throughout the gastrointestinal tract and around the airways of the lung. Contraction of the latter caused narrowing of the airways and an attack of asthma. Histamine was shown to have the opposite action on the muscle forming the walls of blood vessels; when injected into human skin it caused widening of the vessels, seen as a spreading red patch. This is accompanied by a white blister, called a weal, caused by passage of fluid through the blood vessel walls, made leaky by histamine. We know now that histamine is not the only chemical released from the mast cell, and that there are other substances which have a potent effect on blood-vessel walls, the muscle in the lung, and in attracting cells to the site of allergic reaction.

The allergic reaction

There are four components in the allergic response, the basis of all the true allergies seen in man. These are:

● the allergen;
● IgE antibody;
● the mast cell;
● mediators.

The allergens that lead to disease in man are fully described in Chapter 3. These allergens stimulate the immune system of the body causing it to make large quantities of IgE antibody. The mechanisms by which IgE antibody is produced against allergens are described in Appendix A. The surface of mast cells is full of specialized areas known as receptors to which the IgE antibody can fix. There are about 300 000 such sites on each mast cell. When allergen combines with IgE antibody attached to a mast cell, a complicated series of chemical reactions in the cell membrane results. This causes changes in the granules with release of mediators, as shown in Figs 3 and 4.

We know a lot about mast cells. The granules within the cells show a fine structure when magnified 50 000 times, consisting of scrolls, whorls, and a lattice formation. Interaction between mast cells and IgE antibodies leads to the granules becoming soluble, and the fluid, containing potent chemicals, leaks out of the cells, as shown in Fig. 3. Each mast cell contains up to five picograms of histamine

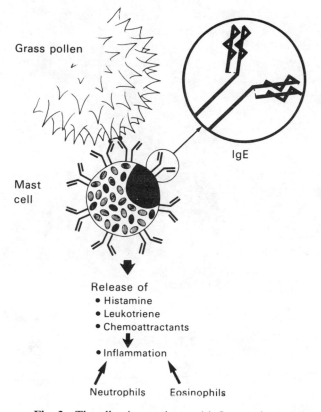

Fig. 3 The allergic reaction and inflammation.

(1 picogram is one millionth of one millionth of one gram), a tiny amount by any standards.

Since the 1930s it has been known that more powerful and longer-acting chemicals are produced during allergic reactions, but it was not until the pioneering work of Vane in Britain and Samuelson in Scandinavia that the nature of these chemicals was elucidated. In the early 1980s, these scientists showed that breakdown of certain components of the cell membranes, known as arachidonic acids, leads to the generation of powerful chemicals. They were awarded the Nobel Prize for Medicine in 1982 for this work. Some of these compounds, called leukotrienes, behave very like histamine but are

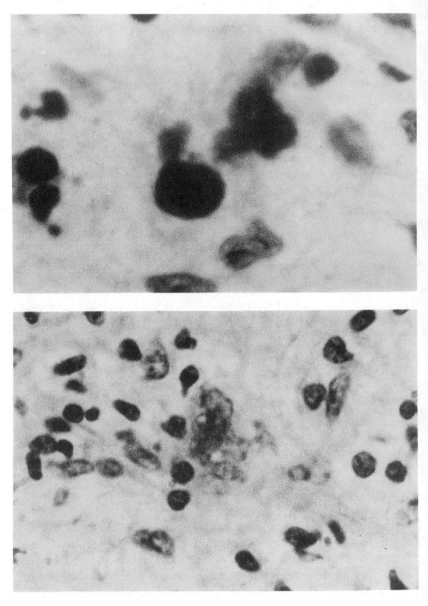

Fig. 4 Mast cells: intact (above) and degranulated (below).

more powerful and have a longer-lasting effect; others attract cells to the site of the allergic reaction.

Defence mechanisms in the body rely not only upon the production of antibodies, which will initially identify and react with proteins present in invading micro-organisms, but also upon cells capable of killing the micro-organisms. The most important of these cells is known as a polymorphonuclear leukocyte (neutrophil) because the nucleus of the cell is many-lobed. These cells form the basis of pus seen in an infected wound, the presence of which is a good sign since it shows there is active rejection of the invading germs. Much modern treatment for cancer involves drugs which not only kill cancer cells but which also kill polymorphonuclear leukocytes. The unfortunate result is that at the critical time of treatment the cancer sufferer may develop severe infections and require large doses of antibiotics until the polymorphonuclear leukocytes once again increase in number. In addition to fighting infection these cells are also attracted to the site of an allergic reaction.

In allergy one type of cell, the eosinophil leukocyte, predominates. This cell has a bilobed nucleus, and like the mast cell contains granules, although not as many. Like mast-cell granules, those in eosinophils contain powerful chemicals, two of which are 'major basic protein' and 'eosinophil cationic protein'. Although these names are complex, the action of these chemicals is straightforward—they cause tissue death and destruction.

Why should eosinophils contain such harmful substances? To answer this it is necessary to go back to the time when invasion by parasites and worms was a major problem to human beings. Even now almost all inhabitants of Third World countries suffer from continuing worm infestation of the gastrointestinal tract. If the worms actually get into the body tissues severe disease and death may ensue. Allergic reactions to these worms are the major source of protection against them. Eosinophils rush to the site of the allergic reaction caused by the worm trying to penetrate the human body tissues. The release of their powerful chemicals either kills the worms or damages them sufficiently to prevent them penetrating the body. Unfortunately, a similar reaction against the harmless allergens from the house dust mite or grass pollen is of no such benefit—quite the opposite. When grass pollen enters the nose or lungs, eosinophils and neutrophils migrate to the site, with subsequent release of damaging chemicals leading to a reaction that is known as inflammation. Basically this inflammatory response, resulting from contact with an allergen, is not too dissimilar to that happening in the skin when bacteria cause a boil, or in lung airways when bacteria cause bronchitis. Often it is extremely difficult to distinguish between the effects of allergy causing allergic rhinitis, and infection with the cold virus (known as rhinoviruses), causing the symptoms of the common cold. Similarly, in the gastrointestinal tract, vomiting, stomach pain, and diarrhoea can result from bacterial infection, viral infection, or food allergy. Allergy is a misplaced inflammatory response. While it is essential to reject invading worms, it is not necessary for the body to react in the same way to harmless allergens from house dust mites, grass pollens, or household pets.

Allergic factors and inflammation

Allergen injected into the skin gives rise to a weal and flare reaction which is largest about 15 minutes after the injection, but careful observation reveals that the reaction is often followed an hour or two later by a further, more diffuse swelling in the skin. This dual response, consisting of an immediate and a late reaction, is due to the effects of initial mast-cell degranulation followed by the subsequent development of inflammation. Exactly the same process goes on in the nose and in the lungs, and probably the gastrointestinal tract, when the allergic reaction is initiated. It is a common experience amongst hay fever sufferers that although they may sneeze when walking through the park on a summer afternoon, worse symptoms of nasal blockage and stuffiness occur in the evening and during the night. Similarly, bakers inhaling flour to which they are allergic during the day often experience coughing and tightness in the chest during the night. What is happening is that the mast cell has been triggered by the allergen to release its powerful chemicals, which attract neutrophils and eosinophils, which in turn cause inflammation and damage in the tissues affected. Since it takes some hours for the cells to be attracted to the site of the allergic reaction and for damage to occur, symptoms come on later during the evening and night.

Examples of immediate and late reactions in the airways of the lungs are fully described in Chapter 6. Their importance cannot be overemphasized. Although bronchodilator drugs that relax the contracted smooth muscle around the airways of the lung work well to counter the effects of histamine and leukotrienes released from mast cells during the immediate response, inflammatory responses involving neutrophils and eosinophils are unaffected, and other treatment, for example corticosteroids, is required. The longer this inflammatory response is allowed to continue following repeated exposure to allergen, the more damaged the tissues become, and the more persistent the symptoms. Studies on sufferers from occupational asthma have shown that, if contact with the allergen is allowed to continue unchecked, then the resulting disease not only becomes more severe but may persist for many years, even when the sufferer leaves the job and is no longer in contact with the allergen that initially caused the disease. This underlines the importance of identifying the allergen, avoiding it as much as possible, and instituting vigorous treatment, particularly with anti-inflammatory drugs such as corticosteroids, at an early stage.

Hyper-responsiveness

Inflamed and irritated tissue at the site of allergic reaction is hyper-responsive: that is, it responds to a whole variety of natural stimuli which previously did not affect the sufferer. Once asthma is established, cold air, cigarette smoke, exhaust fumes, changes in temperature, aerosol sprays, sulphur dioxide, and ozone in the air will all bring on an attack. These stimuli are not acting as allergens but as irritants which are triggering a response in the inflamed and irritated tissue.

Continuing exposure to allergen also causes mast cells to increase in number, making the tissues of the body even more responsive to further contacts with allergen. For example, at the start of the hay fever season in late May, moderate amounts of pollen in the air do not induce symptoms in hay fever sufferers. However, as pollen grains increase in number at the height of the season, symptoms develop, mast cell numbers increase, and subsequently symptoms occur at pollen counts which previously have had no effect. The tissue has become 'primed' to respond to the allergen. The control of mast cell numbers is not fully understood, though we do know that such increases in the number of mast cells can be dampened by continuing treatment with corticosteroids. The value of these drugs in the management of allergic disease is fully described in Chapter 11.

Pseudo-allergic reactions and intolerance

Interaction between allergen and IgE antibody is not the only way in which mast cells can be activated. A number of other stimuli can act directly on mast cells causing them to degranulate and trigger inflammation in predisposed individuals, e.g.:

● exercise;

● environmental irritants, such as ozone and sulphur dioxide;

● drugs and medicines, such as morphine, some anaesthetic agents, and antibiotics.

These reactions cannot be differentiated from true allergic reactions except for the fact that specific IgE antibody against these compounds is absent.

Mention has already been made of intolerance to foods. For example, the absence of lactase, an enzyme in the gastrointestinal tract, means that milk sugar (lactose) cannot be broken down, and diarrhoea results. The patient is *intolerant* to milk rather than being allergic to it. Similarly, aspirin and non-steroidal anti-inflammatory drugs (such as Nurofen used to treat headaches and arthritis), tartrazine, food additives, and preservatives can all trigger reactions in the nose, lungs, and gastrointestinal tract which are indistinguishable from true allergy. However, unlike true allergy the mechanism does not involve specific IgE antibody and mast cells. These compounds cause their effects by modifying the breakdown of components of the cell membrane (arachidonic acid), with subsequent increased production of the leukotrienes. Up to one in 20 patients with asthma and nasal polyps is intolerant to aspirin. What proportion of gastrointestinal disturbances is caused by these compounds is not known.

Other allergic reactions

The major mechanism involved in true allergic reactions and the diseases described in this book is interaction between allergen, IgE antibody, and mast cells, but many other types of allergic reaction can occur, giving rise to a whole range of diseases. For example, tuberculosis is caused by an allergic reaction between lymphocytes and the bacterium causing tuberculosis, *Mycobacterium tuberculosis*.

It is the body's reaction to the bacterium that causes disease and death from tuberculosis. Rheumatic fever and some kidney diseases also result from a continuing allergic reaction to allergens present in a different type of bacterium, *Streptococcus*. Such reactions are also seen against otherwise harmless allergens present in the atmosphere which, when breathed into the lung, can cause 'allergic pneumonia' rather than asthma. Because the reaction affects the air sacs in the lung called the alveoli, it is known as allergic alveolitis. Most cases of allergic alveolitis are related to occupation, but can also occur in the home and are fully described in Chapter 6. They result from the inhalation of:

● mouldy hay, causing farmer's lung;

● contaminated water, causing humidifier fever;

● dust from pigeons and budgerigars, causing bird fancier's lung.

Psychological factors

Many people think that stress or emotional problems are the cause of their allergic disease. In fact:

(1) the psychological make-up of patients suffering from allergic disease is no different from those who do not suffer from such diseases;

(2) not everyone suffering from stress or emotional upset develops asthma or other allergic diseases.

Nevertheless there is little doubt that worry and anxiety can bring on an attack of asthma or rhinitis, in those predisposed. This has been demonstrated in a series of ingenious experiments in which people known to be allergic to flower pollen have developed symptoms on entering a room in which plastic flowers were present. They became anxious on seeing the flowers which they associated with an attack of asthma, and so became wheezy.

The more that is known about the mast cell and, indeed, the connection between the brain and the many tissues and cells in the body, the easier it is to explain the effects of stress. Fine nerve endings go right up to mast cells and have been shown to contain a substance which can cause mast cells to degranulate. At present little is known about the importance of this process in allergic disease, but it could help to explain how stress can cause asthma. Interaction between the brain, the nerves, and the tissues in the development of human disease is an area which is at last receiving the attention it deserves.

3

CAUSES OF ALLERGIC DISEASE

Allergens

An allergen is a substance which can produce an allergic response, and most people associate the term with such substances as pollens, moulds, and foods, which cause sneezing, skin rashes, and wheezing. These substances are not inherently toxic and are generally harmless in the absence of an allergic reaction. In reality, it is not the pollen or food as a whole which initiates a reaction but particular components which are the allergens and take part in the allergic response. Naturally occurring allergens are all large or small proteins. Smaller molecules can act as allergens only after becoming attached, or bound, to proteins already present in the body. Proteins are large molecules composed of amino-acid 'building blocks', and nearly all are soluble in water and therefore able to pass across moist surfaces into the body. Within the range of sizes able to enter the body in this way those at the larger end of the spectrum are known to be better (or worse from the patient's point of view!) at provoking allergic responses. There is no proof that any single amino acid or group of amino acids needs to be present to make a protein an allergen. In fact, at the present time there is still no real answer to the question 'What makes an allergen an allergen?' What is known is that allergenic proteins account for only a small proportion (less than one per cent) of the total weight of a pollen grain. Many of the chemicals making up these crude 'allergens', i.e. the pollen grain or the mould, are not soluble and therefore unable to enter the body to become involved in reactions. Of their total protein content, however, as much as 15–20 per cent may be capable of eliciting allergic symptoms. Of these allergenic proteins, only a few will be involved in producing symptoms. These are called 'major allergens'. The remainder of the 15–20 per cent will comprise 'minor allergens', important for only a small number of patients or of little importance for all of them. It is possible that the minor allergens are responsible for 'cross-reactivity', when a patient allergic to one substance reacts to another related one.

One common feature of all allergens is that prior exposure to them is necessary before the body has the capacity to react. This is because exposure must first result in the production of antibodies which are then involved in future immune responses.

Pollens

Pollens are the most common cause of allergic disease and are responsible for the disease known to most people as 'hay fever'. Unfortunately, the term 'hay fever', which can be traced back to the 1830s, is completely misleading—'hay fever' is not caused by hay, nor does it result in a fever. It is more properly called 'seasonal rhinitis' (rhinitis means inflammation of the nose). Symptoms are due to a seasonal allergy caused by pollen exposure as was first demonstrated by Charles Harrison Blackley in 1873.

Pollen grains contain the male sexual cells of plants which fertilize egg cells to produce seed. They are transferred from one plant to another by the wind or on the bodies of insects. Most dangerous for sufferers are the wind-pollinated plants. These plants release vast amounts of pollen grains into the atmosphere. However, a plant whose pollen is wind-borne and allergenic will only be an important cause of symptoms if it is sufficiently widely distributed. Three types of plant are major causes of pollen allergy:

- grasses
- trees
- weeds

Grass pollens are the most important cause of symptoms in Western Europe, birch pollen is the prime offender in Scandinavia, and ragweed in the USA. In Britain the three categories of pollen reach a peak in the atmosphere at different times of the year as illustrated in Fig. 5.

Grass pollen

Grass is the most common cause of seasonal allergy occurring in mid-summer. Amongst the hundreds of species of grasses only a relatively small number are sufficiently widespread and prolific pollen producers to give high atmospheric pollen counts. Some important species are:

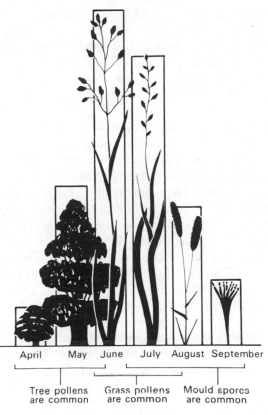

Fig. 5 Histogram of seasonal rhinitis causes by month.

- sweet vernal grass
- orchard grass
- cocksfoot
- rye grass
- Yorkshire fog
- meadow grass
- timothy grass

There is extensive cross-reactivity between these species so that a patient allergic to one type may react to some extent to others. The grass pollen season in Britain extends from mid to late May through

to early August, peaking in June and early July. Within this season
the amount of pollen in the atmosphere varies from day to day and
from hour to hour. For instance, each species of meadow grass sheds
its pollen at its own specific time of day. Most species flower once
daily in the early morning. However, two of the most prolific pollen
producers, Yorkshire fog and sweet vernal grass, flower for a second
time in the late afternoon.

As grass flowers open some pollen will be scattered immediately,
but a considerable amount will be deposited on nearby leaves, etc.
and will only be blown away later by stronger gusts of wind. The
amount of pollen in the air closely relates to the weather, being
affected by both temperature and humidity. On fine, clear days there
will be a high concentration of pollen in the air and pollen will be
carried up through the atmosphere by warm air currents, sometimes
to the height of cumulus clouds. When these convection currents
cease as the Earth's surface cools in the early evening, pollen grains
fall to form concentrated clouds. At this time, between about 5 and
6 p.m., the peak concentration of pollen occurs in country areas. In
central London the peak is delayed by 1–2 hours—the time it takes
for pollen clouds to be blown into the city centre. On calm nights
pollen from higher in the atmosphere will gently settle out in the still
air, resulting in a second peak after midnight. This pattern can be
disrupted by rain in the morning, delaying the opening of the grass
flowers, and, if continuous, reducing pollen counts that day. However, if rain occurs in the afternoon when the pollen concentration is
rising, the air will be cleared leaving sufferers a sneeze-free evening!
Daily estimates of the pollen count (expressed as the average
number of grains per cubic metre of air during a 24-hour period) are
now regularly reported in newspapers and on television and radio
news broadcasts. At the beginning of the grass pollen season, hay
fever sufferers usually begin to experience problems when the pollen
count reaches about 50. However, recent research has shown that as
the season progresses and the nose becomes increasingly irritated,
less and less pollen is required to produce symptoms. This phenomenon is known as 'nasal priming'.

Tree pollen

Pollen is produced by many deciduous trees during the spring,
shortly after the leaves develop. The pollen season tends to be of

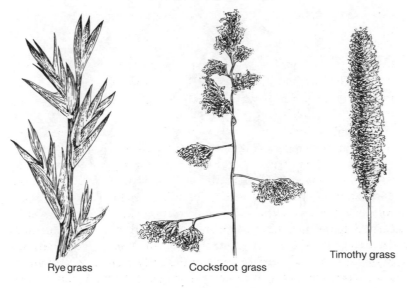

Rye grass   Cocksfoot grass

Timothy grass

Fig. 6   Allergenic grasses.

shorter duration and more clearly defined than that of grass pollens. In Scandinavia, birch pollen is the major allergen. In general, symptoms in the UK are caused by oak, elm, birch, plane, ash, and hazel. Whereas pollen from oak and elm may be important, that from chestnut and beech is rarely implicated. Other trees are important in different countries, e.g.:

- birch in Scandinavia;
- olive in Mediterranean countries;
- cedar in Japan;
- mountain cedar in the USA.

Conifers, for example pine trees, are prolific pollen producers but do not generally cause rhinitis since the pollen is non-allergenic.

Weed pollens
Weed pollens are of less importance in Britain and Europe than in the USA, where six cross-reacting species of ragweed (a relative of the mugwort) are a major problem. Ragweeds grow on sand-bars,

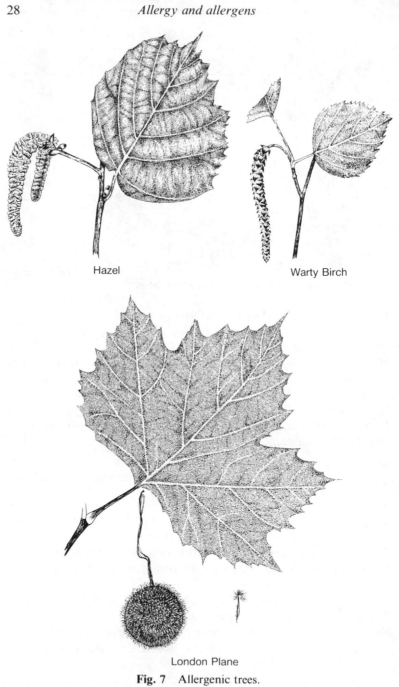

Hazel

Warty Birch

London Plane

Fig. 7 Allergenic trees.

floodplains, roadsides, and especially in the grain fields of the Midwest. Ragweed plants, which are prolific pollen producers, shed their pollen over a short period of two weeks in August or September. One square mile of ragweed plants produces 16 tons of pollen! It is hardly surprising that rhinitis in the autumn is such a widespread and severe problem in the USA.

Stinging nettle   Nettle leaves are covered with stinging hairs which inject histamine-releasing substances into the skin, causing the typical pain, itching, and rash. Nettles flower in late summer and autumn, ejecting pollen into the air as the sun shines on the curled young flowers. The importance of nettle pollen in autumn-time allergic symptoms is not really known.

Brassicus napus—oil-seed rape   Over the last few years the cultivation in the United Kingdom of oil-seed rape has been increasing. In Scandinavia, where the cultivation of oil-seed rape increased considerably in 1939 to supply the demands of the margarine industry, an increased incidence of rape pollen allergy has been observed over the past five years. In a study of 366 consecutive patients suffering from diagnoses of asthma, pollen allergy, vasomotor rhinitis, bronchitis, urticaria, and suspected allergy, or a combination of these conditions, 23 per cent were found to have positive intra-cutaneous skin tests to rape pollen. There are different pollinating seasons in Scandinavia for winter turnip rape, spring turnip rape, winter rape, and spring rape; all between 10 May and 24 June.

Recent newspaper reports have suggested that racehorses living in the vicinity of oil-seed rape have diminished performance and general lassitude, which improves after the rape crop has been harvested.

Wall pellitory—Parietaria officinalis   This is a common weed, particularly in Mediterranean countries, though it is also found in the United Kingdom. It is perennial and grows on walls, rocks, and banks. Although recognized as an important cause of allergy in southern Europe, few cases have been reported in England. Skin-testing of allergic people living in the Southampton area where the weed is common have shown that less than 10 per cent have positive tests.

House dust and the mite

Any dust can be irritant if inhaled in high concentrations. Everyone has experienced sneezing when clearing out cluttered garages or neglected attics. However, for some people dust is more than an irritant. For people who become sensitized to a constituent of household dust, exposure to relatively small amounts can trigger an attack of sneezing or breathlessness.

Allergy to house dust is common especially in those who have symptoms all year round. In Britain over 80 per cent of asthmatic children show skin reactions to house dust extracts. House dust is composed of a variety of substances, including moulds and skin scales shed from animals, but a major constituent is the house dust mite and its produces.

The house dust mite

In 1928 a German scientist called Dekker suggested that microscopic mites living in dust from bedding could be a major cause of asthma. This idea was taken no further until the 1960s when techniques for recovering mites from specimens of house dust had been developed. The house dust mite is now acknowledged as the major source of allergen in house dust in many countries. In Britain the dominant species is *Dermatophagoides pteronyssinus* (from the Greek meaning 'skin-eating feather mite'), accounting for 74 per cent of the mite population in samples obtained from London homes. Of less importance in Britain is a related mite, *Dermatophagoides farinae*, which, although representing only about three per cent of the mite population here, is the dominant species in the USA. As Dekker first reported, *Dermatophagoides* mites are 'bed mites' which tend to be concentrated in highest numbers in and around beds. The reason for this is that they feed on human skin scales which are shed mainly when in bed. There is no shortage of food, since one person sheds 0.5–1 g of scales each day—sufficient to feed thousands of mites for months. Since their diet of skin scales is essentially dry, mites need to obtain fluid from the air, and for this reason the humidity of the atmosphere is usually the decisive factor in determining the size of the mite population. Conditions of approximately 80 per cent relative humidity and 25 °C of warmth form an ideal habitat for mites. Since in London the average monthly humidity varies between about 73 per cent in June to about 86 per cent in December, it is not surprising that levels of infestation in British homes are high. The increased prevalence of mites in humid areas is emphasized by the fact that in London mite numbers have been shown to correlate with distance from the Thames. In addition, the homes of allergic people are often found to be situated close to underground waterways, as shown in Fig. 8. In mountainous areas, such as the Alps, mites are found in very low numbers, their scarcity probably relating to low temperatures in conjunction with low humidity.

Levels of humidity within the home do not depend solely on local climate; the humidity within a bed and bedroom also depends on:

(1) the number of people sleeping in the room (each person loses 500 ml of water in evaporation during the course of one night);

(2) the size of the room;

(3) the degree of insulation;

(4) frequency of airing of the room;

(5) the presence of a humidifier.

Due to the factors mentioned above, mites vary in numbers between homes from 0 to 2000 per gram of surface dust. Being approximately 0.3 mm in size, mites are indistinguishable from specks of dust to the naked eye. This is perhaps fortunate since high magnification shows them to be of unprepossessing appearance!

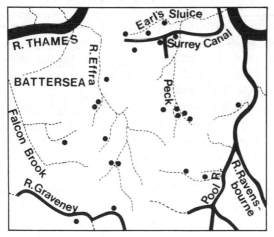

Buried waterways---- Open waterways ━━━
Houses of allergic cases ●

Fig. 8 Map of allergy cases and underground rivers.

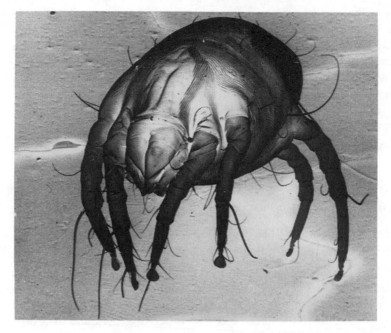

Fig. 9   House dust mite.

(Fig. 9). Despite their small size, intact house dust mites are far too large to be inhaled deep into the nose or lungs. Attention has therefore centred on smaller fragments of mite as the potential allergen. A substance (called Der p1), found in particularly high concentrations in mite faecal pellets, has now been identified as the major allergen. Unfortunately for sufferers, mite faeces are of a size comparable to pollen grains, and can be inhaled easily into the nose, leading to allergic rhinitis. Even the small size of the mite faecal pellet (20 microns, i.e. twenty-thousandths of a millimetre) is too big to be inhaled directly into the lungs, and it is likely that allergic asthma arises from inhalation of dried fragments consisting of the Der p1 allergen.

Storage mites

Closely related to the pyrogliphid mites, such as the house dust mite, are the acarid mites, such as *Acarus siro*, *Glycyphagus destructor*, and *Glycyphagus domesticus*. These species are storage mites commonly found in granaries, warehouses, food stores, and farm stores. These mites are even more sensitive to desiccation than the house dust mites and need a very moist atmosphere and prefer very warm temperatures (in the range 25–30 °C). Unlike house dust mites, storage mites prefer a vegetarian diet of grain and flour and are only infrequently found in the domestic environment. However, they do often outnumber house dust mites in dust from living accommodation in the tropics. This is partly due to the high humidity, but is perhaps even more attributable to the fact that the same room is often used for cooking, sleeping, and food storage.

Domesticated animals

Throughout history man has had a close association with animals, both at work and in domestic life, but contact with them inevitably leads to exposure to allergens in the form of proteins in hair, skin scales, saliva, and urine. Many people experience no ill effects from this exposure, but for the allergic patient it can be the cause of distressing symptoms. Animals frequently reported as the cause of allergic reactions are:

● cats
● dogs
● horses
● small mammals, such as rats, mice, and hamsters
● cage birds

Cats and dogs

These are most frequently encountered as pets in the domestic environment, and up to 30 per cent of allergic patients report reactions to these animals. The most important cause of symptoms is allergens in skin scales, known as dander. Cat and dog hair is a much poorer source of allergen, although it may appear to provoke symptoms since dander is often adsorbed to the surface of the hair after it has been shed from the skin. Proteins in urine and saliva can also cause symptoms, but are of less importance in the development of allergic symptoms. Direct contact with a cat or dog is not essential for allergic symptoms to develop, since the whole environment of a household where pets are kept will be contaminated. Dander is an important constituent of household dust and, in patients with perennial symptoms, it is important to differentiate between symptoms due to the house dust mite or to animal danders.

Patients often claim that certain breeds of cat or dog cause more problems than others. It is unlikely that allergens are breed-specific but other explanations are possible:

(1)   some shorter-haired dogs shed more dander than long-haired breeds;

(2)   the amount, rather than the type, of allergen present in dander may vary between breeds.

Allergens derived from different types of cat have not been thoroughly studied, but it is possible that distinctive allergens occur in Siamese cats.

Horses

Horses can be a cause of allergic disease in a surprising number of people including:

● agricultural workers

● stable/racecourse workers

● people who ride as a hobby

Horse dander is the major cause of symptoms. Indirect exposure to horse allergens has fallen with the decreasing use of horsehair in the manufacture of furniture and mattresses but can still be implicated in disease. Horse riders and their families, for example, are in indirect contact through horse allergens brought into the home on clothes: this may be sufficient exposure to precipitate symptoms in highly sensitive individuals. Even more obscure forms of exposure do occur, for instance in the case of gardeners exposed to horse manure, or in zoo attendants, since there is a possibility of cross-reactivity with mules, donkeys, and zebras. Cross-reactivity also occurs between horse dander and horse serum, which may be important in the light of the fact that some vaccines (e.g. tetanus) are prepared using horse serum.

Small mammals

Small mammals including mice, guinea pigs, hamsters, and gerbils are all potential sources of allergic sensitization when kept as pets. In addition, medical research workers may be occupationally exposed to high levels of allergenic materials from rats and other small mammals. Urine is a more potent source of allergen than dander in these animals, and material lining cages will be heavily contaminated. Urinary proteins are released into the air as the animal moves around the cage. Up to 20 per cent of laboratory workers involved in handling small mammals become sensitized, with symptoms usually developing within one year of their first exposure. Some employees with an allergic background are at increased risk of sensitization, though this problem may eventually affect all those employed.

Birds

Birds can give rise to allergic disease in several ways. Feathers, especially in the form of pillows or feather beds, are known to cause symptoms in many patients. However, reactions appear to occur much less frequently to freshly plucked feathers than to those which have been stored. This suggests that the allergenic substances are degradation products of feathers. In fact, these reactions are due to contamination by house dust mites. Patients who experience worsening of their symptoms after sleeping with feather pillows or duvets are usually allergic to house dust mites. Allergy to bird droppings also occurs. This is unlikely to develop from feeding pigeons in the park, but does occur in pigeon fanciers, who have heavy exposure in poorly ventilated coops, lofts, and sheds. Occasionally, allergic symptoms can arise from the dried droppings present in bird cages. Luckily this is very rare, since caged birds, particularly budgerigars, are kept in up to a quarter of homes. When inhaled, proteins from dried droppings can cause additional diseases, particularly in the lungs, giving rise to scarring and severe breathlessness.

Mould

Moulds are microscopic living organisms, composed of long filaments, which usually grow on the surface of organic materials such as compost heaps or rotten fruit and cheese. They are classified as a sub-division of the fungi. Moulds are very adaptable and may be found in almost any place where there is moisture and oxygen. Optimal conditions for mould growth occur in warm weather with relatively high humidity. In forests moulds grow on rotting logs and vegetation, especially in moist, shady areas. In the home they are often found in damp basements, bathrooms, refrigerators, and upholstered furniture.

Moulds have life cycles of varying complexity, producing reproductive bodies (spores) which are dispersed, in most cases, by air currents. The concentration of spores in the air, and the species of mould from which they originate, depends on many factors, e.g.

● temperature

● atmospheric moisture

● season

● prevailing winds

Some spores are propelled from the mould into the atmosphere by processes which depend on the presence of water, and therefore increase during periods of dampness and rainfall, whereas others are blown free by the wind and increase during times of low humidity and increased air flow.

Moulds and their products have been of great benefit, not only for improving the flavour of certain cheeses and wines but, more importantly, in forming the basis of nearly all antibiotics, from penicillin onwards.

Even though it has been known for over a century that inhalation of mould spores can produce symptoms, the importance of their role in allergic disease remains controversial. There are many reasons for this:

(1)   difficulties in identifying individual moulds;

(2)   the absence of a clear seasonal pattern in spore production.

Many moulds are commonly encountered.

Cladosporium herbarum

Spore production by this mould is prolific, with concentrations at the height of the season exceeding 10 000 per cubic metre. For this reason it is the most commonly encountered mould allergen. The spores are present in the atmosphere throughout the year, with heavy concentrations from May to October, especially in the centre of cities (up to 15 000 spores per cubic metre) from mid-June to mid-September. This mould commonly colonizes foliage and plant matter, particularly grass. Airborne spores increase dramatically when the lawn is mowed or trees are pruned. The mould is found in dirty refrigerators, on moist window frames, and in houses situated in low, damp environments with poor ventilation. This mould often contributes to 'hay fever', especially in cases where symptoms do not appear to coincide with the grass pollen count.

Penicillium notatum

This is the best-known penicillin-producing mould and is very widely distributed. In the domestic environment *Penicillium* is the blue-green mould found on stale bread and citrus fruits and its spores are frequently found in household dust. For some reason the spores of the *Penicillium* group appear to be more abundant in town compared to rural areas. It has no marked seasonal variation but reaches peak concentrations in the winter and spring.

Alternaria alternata

This is an extremely common mould found in soil, compost, and rotten wood, but also in textiles and even foodstuffs. Black spots on tomatoes are often caused by this mould. *Alternaria* is also frequently present on window frames where condensation occurs. In general, it is considered to be an outdoor mould, appearing during warm weather. It is one of the most important mould allergens in the USA and, perhaps for that reason, has been extensively researched. In fact it is one of the few substances whose major allergen has been identified; it is named Alt 1. *Alternaria alternata* also has some effects not related to its involvement in allergic disease. For example, heavily contaminated cereal foods can be directly toxic to warm-blooded animals.

Botrytis cinerea

Botrytis cinerea is called the 'grey mould' as it covers decayed tissues with grey-brownish spores. It is seen especially in connection with soft fruits, vegetable crops, seedlings, and young plants. Horticulturists commonly encounter this mould. The growth of *Botrytis* on grapes is regarded as enhancing the bouquet of certain wines, e.g. Chateau d'Yquem, and has been reported as the cause of allergic symptoms in vineyard workers. Spores, which are shed between noon and 7 p.m., occur in the atmosphere from the beginning of

May to the end of December, with highest concentrations from mid-June to mid-October.

Aspergillus fumigatus

This mould is distributed world-wide largely because it can grow at any temperature between 12 and 57 °C. In spite of this, the concentration of spores in the atmosphere is low in comparison to other airborne allergens, such as pollen. *Aspergillus fumigatus* is unique amongst moulds causing allergic symptoms, in that it can actually grow in the lungs of patients who have normal immune defences. This gives rise to a variety of diseases:

(1) allergic pneumonia;

(2) damage to the airways of the lung, a condition called bronchiectasis;

(3) aspergilloma, a ball of mould hyphae growing in cavities in the lung.

A mould closely related to *Aspergillus fumigatus* is *Aspergillus clavatus*, the spores of which are responsible for a common respiratory disease of malt workers.

Candida albicans

Candida albicans is another mould which, like *Aspergillus*, is capable of colonizing the human body. *Candida* is a white or cream-coloured yeast with a slimy appearance which grows on the moist surfaces of the nose, mouth, and throat of babies, causing thrush. In adults it may also occur in the mouth, though it is rare. It is more frequently responsible for vaginal infections or for skin infections in diabetic patients. Although *Candida* is also common in soil and organic debris, the spores do not easily become airborne. For this reason its role as a cause of allergy has been disputed, but allergy to *Candida* has been reported in patients with asthma, sneezing, and urticaria (a type of allergic skin rash).

Didymella exitalis

This is a mould which is found in many countries on the leaves of ripening barley and wheat. Spores first appear in the air in mid-June and are still present in late August and early September. In contrast to *Cladosporium* and *Alternaria*, where high spore counts occur on dry summer days, the spores of *Didymella* require moisture for their release. This moisture is normally provided by dew formation, causing spore release to begin around midnight, reaching a peak at about 3 a.m. However, rainfall can disturb this pattern, causing release at other times; the number of airborne spores reaches a maximum 4–6 hours after the rain has stopped. The effect of thunderstorms is dramatic, increasing spore counts from a daily mean of below 1000 spores per cubic metre of air to more than 25 000 about 6 hours after the start of the rainfall. Allergy to *Didymella* probably accounts for the strange occurrence of 'epidemics' of asthma following thunderstorms, which have been reported both in Britain and in Spain.

Insects

Insects are one of the most successful and numerous classes in the animal kingdom. They are encountered in almost all environments, often in extremely large numbers, and perhaps it is not surprising, therefore, that they have a role in allergic disease. Sensitivity to insects falls into two categories:

● inhalant insect allergy
● allergy to insect stings/venom

Inhalant insect allergy

Parts of insects are often seen as a constituent of house dust on microscopic examination or when sampling the atmosphere for airborne allergens. However, it is the smaller particles of insect debris, such as shed skin scales, dried secretions, and faecal particles, which may cause symptoms of asthma or rhinitis when inhaled by sensitized individuals. Many insects including:

● aphids
● cockroaches
● locusts
● mayflies
● green nimitti flies

can be implicated in this type of allergic disease. A good example of this type of insect allergy is sensitivity to the green nimitti midge

(*Chironomus lewisi*). These midges are small, non-biting flies between 2 and 15 mm in length. They are characteristically seen swarming by water at dusk. Chironomids occur in nuisance numbers world-wide but are a major cause of allergic symptoms in the Sudan, where sensitivity occurs due to inhalation of small fragments of the insect body. It has been shown that the important allergens are two types of haemoglobin (the oxygen-carrying protein) found in many different chironomid species. This means that people becoming allergic to one particular species will react to others: for example those sensitized to *C. lewisi* will also react to *C. riparius*, whose larvae is used in Europe as pet-fish food. Attacks of rhinitis due to sensitivity to these allergens has occurred in people exposed to this food as part of their employment. (Symptoms have also been known to result from inhalation of particles of dried *Daphnia* used as food for ornamental fish.) This type of obscure exposure makes insect allergy difficult to diagnose.

Insect sting allergy

Almost everyone is aware of the unpleasant effects of an insect sting. Such insects belong to the order Hymenoptera which has two families:

- Apoidea—honeybees and bumble-bees
- Vespoidea—wasps, yellow jackets, and hornets

The honey and bumble-bee tend to sting only when provoked; they die in the process of stinging. Stinging usually occurs in defence of the hive and therefore beekeepers are at increased risk. Wasps, yellow jackets, and hornets can sting repeatedly, using their venom to paralyse and kill insects to feed their larvae. They are aggressive and may sting even when unprovoked.

Hymenoptera venoms are complex mixtures of substances, some of which act as toxins and others, the protein components, which act as allergens in sensitized people. The toxins present in the venom cause pain, redness, and swelling lasting for 1–2 days as a *normal* reaction to stings. Large reactions lasting for several days can also occur as the result of infection. Such large reactions, however, may be due to allergy. Any generalized reactions not arising directly from the area of the sting are almost certainly allergic. Only in rare cases of multiple stings is any generalized effect on the body directly attributable to injected toxins.

Diagnosis and treatment of allergy to stinging insects is complicated by the difficulty of identifying the offending insect. Bee stings are more readily identifiable since the sting is left in place, but it may be almost impossible to differentiate between the various types of vespid. Further confusion arises from the fact that patients may be individually sensitive to more than one insect or may react to an allergen shared by the different insects to which they are allergic. For example, allergenic substances common to both the wasp and yellow jacket have been identified. Ideally, bee stings should be removed by holding the sting (preferably with tweezers) as close to the skin as possible to avoid squeezing more venom into the wound.

Foods

The word 'allergy' is frequently over-used and perhaps never more so than in connection with abnormal reactions to foods. In fact adverse reactions to foods can occur by allergic or non-allergic mechanisms resulting in:

- food allergy
- food intolerance

Food intolerance can arise for a variety of reasons. Reactions may occur by non-immunological mechanisms, be due to poor processing of the food in the body as a result of an enzyme deficiency, or arise from the presence of histamine or histamine-releasing agents in the food. Histamine is capable of causing symptoms normally attributed to allergy and can occur in surprising amounts in foods. For example, old cheese can contain 85 milligrams per 100 grams of food and tuna fish 400 milligrams. Foods commonly implicated in food intolerance or allergy, either singly or in combination, include:

- cow's milk and milk products
- goat's milk
- hen's eggs
- colouring agents
- preservatives
- salicylate-containing foods
- cereals, including wheat and oats

- yeast
- fish, including shellfish
- nuts
- chocolate
- tea and coffee
- pork, bacon, and tenderized meats
- soya
- tomatoes
- peas
- beef
- chicken

This list is by no means exhaustive.

The components of foods which cause allergic symptoms are durable and often survive cooking or digestion in a form still capable of causing a reaction. Almost any food can cause allergic symptoms, but some are more allergenic than others. In some food groups an allergy to one member may result in a person being allergic to other foods in the same group (a phenomenon known as 'cross-reactivity'). However, within groups of animal foodstuffs cross-reactivity is rare. For example, people allergic to cows' milk can usually eat beef.

Cow's milk
Cow's milk is a prime offender in infancy and remains a major cause of allergy in later childhood. There is often cross-reactivity between cow's and goat's milk, since these animals are related. Cow's milk allergy is a controversial subject since immune reactions can be demonstrated in only a small proportion of patients who report symptoms. As described earlier, many people may be suffering from intolerance rather than allergy. This may be due to a lack of the enzyme lactase which means that lactose present in milk is neither digested nor absorbed. In some cases there are more obscure reasons for milk intolerance. For example, traces of penicillin in cow's milk can cause symptoms in patients highly sensitive to this drug.

Eggs
Eggs are of great importance in food allergy. The most allergenic

part of the egg is the egg-white protein, and reactions to egg yolk are usually due to contamination with white. People allergic to egg usually tolerate chicken meat but may react to vaccines grown on egg.

Fish

The first pure allergen ever identified was isolated from cod muscle. Fish allergens are very potent; even inhalation of steam from cooking fish can cause asthma in very sensitive patients. Of those allergic to fish, 50 per cent react to all species of bony fish, although others can tolerate some fish.

Shellfish

Shellfish in general, especially shrimps, lobster, and crab, have a tendency to cause violent allergic reactions. However, sensitivity to shellfish does not imply cross-reaction with bony fish.

Nuts

Nuts often identified as the cause of allergic reactions include:

● peanut (ground nut)
● cashew nuts
● hazel-nuts

In people with symptoms due to tree pollen allergy, reaction to walnuts and hazel-nuts is common.

Fruit and vegetables

Sensitivity to fruit and vegetables is common, especially so in patients with birch pollen allergy who often react to various combinations of the following:

● apple
● peach
● cherry
● pear
● carrot

Cereals

Allergic reactions can occur as a result of products made from various cereals including:

- wheat
- rye
- barley
- oat
- corn
- rice

There is cross-reactivity between wheat, rye, and barley, which all belong to the same family. A reaction to rye products does not relate to seasonal ryegrass allergy since in the latter it is the pollen rather than the grain which is responsible for symptoms. However, inhalation of grain fragments can result in occupational disease such as baker's asthma. In addition to allergic reactions, many people are intolerant to gluten, the protein which gives dough its tough, elastic character, and forms about 10 per cent of wheat flour. This is discussed more fully in Chapter 8.

Soya bean
Soya protein is often used as a substitute for other proteins, e.g. cow's milk. It is a weak allergen only and sensitization is rare.

Alcoholic drinks
Alcoholic drinks often precipitate symptoms in allergic individuals. This is due to the range of dyes, preservatives, and congeners which such drinks contain.

Food additives
Food additives include chemicals added to food as preservatives, flavourings, and colouring agents. Less than 1 per cent of the general population are intolerant to additives, suffering symptoms of an allergic-like nature. Probably only a proportion of such reactions are truly allergic. The yellow dye tartrazine (E102), used to colour both food products and some drugs, may be particularly associated with problems in patients intolerant to aspirin. Another group of chemicals to which people may become sensitized are the sulphiting agents. These are used to eliminate bacteria, preserve freshness and brightness, increase storage life, and prevent spoilage of some food products. Sulphites are also used in the sterilization of home-brewing and wine-making equipment; and in the preservation of

grapes before pressing so that levels in certain wines may be extremely high. A more comprehensive list of chemicals occurring as food additives is included, with advice on avoidance, in Chapter 10. Finally, remember that all cosmetics, toiletries, toothpaste, and medications must also be considered as possible allergen sources.

Occupational allergens

In a surprising number of cases 'allergic' problems reported by patients are related to materials encountered in the workplace. In Japan, for example, up to 15 per cent of all male cases of asthma are thought to result from exposure to industrial gases, vapours, or fumes. Even the common environmental allergens already discussed, such as pollens or animal danders, may be important occupational allergens to people such as gardeners, florists, and laboratory workers.

Occupational disease was first reported by Paracelsus in the sixteenth century but the 'father' of occupational medicine is Bernadino Ramazzini, who was a Professor of Medicine in Padua. He stressed the importance of a patient's occupation in relation to disease in his book *Diseases of workers* published in 1713. Ramazzini focused attention on breathing problems and skin rashes occurring in grain workers, and stressed the importance of recording the type of work in which each patient was involved. In more modern times occupational diseases are widely recognized. Not all occupational materials produce true allergic reactions; some are toxic, some are simply irritant, and some are capable of mimicking an allergic reaction by causing histamine release in the body without involving the immune system.

Agents involved in occupational disease are many and varied. One of the most important groups of chemicals causing asthma is the isocyanates, particularly toluene diisocyanate (TDI), which are widely used in the manufacture of polyurethanes. Respiratory problems related to exposure to TDI were first reported in 1951, when nine out of 12 workers in a chemical factory in France were found to have asthma. Exposure to TDI increased dramatically with the development of the plastics industry. TDI is an irritant material but minute quantities can also cause sensitization and 'allergic' symptoms. This underlines the difficulties of separating symptoms resulting from irritation from those due to allergy. TDI vapour may act as an irritant in some workers while sensitizing others.

Some of the commonest materials encountered at work which are associated with occupational disease are shown in Table 1.

Table 1

Material	Industry	Mechanism of action
Ammonia	Chemical and	Irritant
Sulphur dioxide	petroleum	
Chlorine	industry	
Nitrogen dioxide	Silo fillers	
Flour	Baking	Allergic
Grain	Baking, farming	
	Grain terminal workers	Allergic
Enzymes from *Bacillus subtilis*	Detergent manufacture	Allergic
Castor bean	Food processing	Allergic
Green coffee bean		
Phthallic anhydride	Plastics industry	Allergic
Trimellitic anhydride		
Complex salts of platinum	Metal refining	Allergic
Wood dusts	Saw mills, carpentry	?Allergic
Penicillin	Pharmaceutical	?Allergic
Sulphonamides	industry	
Soldering fluxes	Electronics industry	?Allergic/irritant
Toluene diisocyanate (TDI)	Paint, plastics, rubber, and resin industry	?Allergic/irritant/toxic
Cotton dusts	Textile industry	Toxic/?allergic

Isocyanates

Since the 1940s there has been a huge increase in the use of isocyanates in industry, where they are used in the production of polyurethanes for the manufacture of:

● man-made fibres
● adhesives

- insulation
- paints/varnishes
- foams (flexible and rigid)

The isocyanates are an extremely reactive family of chemicals, of which the most important are:

- toluene diisocyanate (TDI)
- diphenylmethane diisocyanate (MDI)

TDI is the more widely used compound, its major application being in the manufacture of foams. High concentrations of vapour of more than 2 parts per million (p.p.m.), such as may occur in major accidental spillage, or when isocyanate-containing material is on fire, have a direct irritant effect which causes breathing difficulties in everyone exposed. However, many workers become sensitized to much lower concentrations of less than 0.02 p.p.m. About 5 per cent of workers become sensitized to TDI following exposure, which may occur in various situations, including:

(1)   manufacture of the chemical;

(2)   leakage during bulk handling of TDI;

(3)   disposal of TDI waste;

(4)   spraying and moulding operations;

(5)   welding polyurethane-covered wires.

TDI may also be encountered away from the work environment, for example, when polyurethane varnishes are used with a TDI activator or when polyurethane products are burned.

Platinum salts

Complex salts of the metal platinum, such as ammonium hexa-chloro-platinate and potassium tetrachloro-platinate, are used in metal refining and in the photographic industry. Their harmful effects on the respiratory systems of photographic workers in Chicago was first reported in 1911. The importance of this finding was only recognized years later, in 1945, when the frequent occurrence of sneezing and wheezing amongst London metal-refinery workers exposed to the complex salts of platinum was reported.

Extremely small quantities are often sufficient to cause attacks of wheezing and breathlessness, indicating that reactions to complex salts of platinum are allergic in nature.

Enzymes from bacteria

In the 1960s enzymes from *Bacillus subtilis* were added to detergents to produce the first 'biological' washing powders. By 1969 it became clear that some employees involved in the production of these powders were likely to develop asthma. Reactions to the enzymes alkalase and maxatase are now well recognized. Originally, up to 2 per cent of exposed workers developed symptoms due to the inhalation of enzymes from *Bacillus subtilis*. More recently, better control of the industrial environment has caused the incidence to fall. Use of biological washing powders at home is not thought to involve any risk since the enzymes are enclosed in starch, making the particle far too big to be inhaled into the respiratory tract during the normal process of pouring the powder. Many people, even those without rhinitis, notice that pouring washing powder causes sneezing. This is not due to an allergy but the result of the irritant nature of the fine particles of soap which are strongly alkaline and are readily breathed into the nose.

Solder flux (colophony)

In the past as many as one-fifth of solderers in some electronics factories developed severe respiratory symptoms, initially diagnosed as chronic bronchitis. In fact, the symptoms resulted from the inhalation of the fumes of colophony, a pine resin, used as soldering flux, which had led to the development of asthma. Fluxes are widely used in the electronics and related industries, and workers may become sensitized to one type of colophony, for example Portuguese, but not another, for example American. In some sensitized workers symptoms persist even when they are no longer exposed to fluxes at their place of work. It is possible that this is due to the use of colophony esters and amber oil for the flavouring of cigarette tobacco. In addition, sensitized individuals may develop attacks of sneezing and wheezing when walking through pine woods or from the Christmas tree in the home.

Wood dusts

Some hardwoods are being used in increasing quantities in both indoor and outdoor construction work. One of the most widely used is Western red cedar (*Thuja plicata*), the dust of which has been shown to cause asthma in some workers. One of the substances present in the dust and responsible for the development of symptoms is the non-volatile plicatic acid. Other woods commonly implicated in allergic disease are:

- Cedar of Lebanon
- Iroko

Grain and flour dust

Asthma associated with grain dust and flour has been recognized since Ramazzini first published his observations on the subject in 1713. People likely to be exposed to grain dust or flour in the course of their work are:

(1) farm workers during harvest time and when handling stored grain in the winter;

(2) dockers and grain elevator operators;

(3) millers;

(4) bakers and pastry cooks (baker's asthma).

Grain dust is a composite mixture of particles from various cereals, such as wheat, barley, oats, rye, and maize. In addition, there is often contamination from a variety of sources such as:

- pollens and seeds
- bacteria
- fungi
- insects, such as grain weevils
- mites
- chemical pesticides and herbicides

About 40 per cent of the airborne dust from grain consists of particles small enough to enter the human respiratory tract. The composition of grain dust itself varies according to the locality and climatic conditions in which the grain is grown, transported, and stored.

Grain dust produced during harvesting Cereal plants are often infected by fungi with airborne spores. Dust generated by combine harvesters will therefore contain enormous numbers of fungal spores. The most frequently encountered mould is *Cladosporium*, but others such as *Verticillium, Alternaria, Paecilomyces*, and *Ustilago* are also common.

Dust produced by stored grain   The composition of airborne dust from stored grain varies greatly. For example, the nature of the contaminants growing in the grain (and their airborne spores) is determined by storage conditions. As the water content increases to about 30 per cent and the temperature to between 50 and 65 °C, greater numbers of organisms thrive in the grain. Sealed silos with little air available limit the spontaneous heating of the grain, keeping the temperature below this optimum level. However, maltings, where the grain is aerated, encourage the growth of moulds such as *Aspergillus*. In addition to the fungi, whose spores may result in allergic disease, bacteria are always present in grain. Toxins produced by these bacteria may be responsible for non-allergic respiratory diseases. Stored grain is often infested with mites, closely related to the mites found in house dust. In the UK some of the most commonly encountered storage mites are:

● *Acarus siro*

● *Tyrophagus putrescentiae*

● *Glycyphagus domesticus*

These mites cause occupational asthma in the same way as the house dust mites.

Antibiotics

Antibiotics of the penicillin type can cause allergic reactions during courses of treatment and perhaps it is not surprising that workers involved in their manufacture should also develop allergic disease. No single substance contained in these drugs is responsible for the allergic reactions: different workers react to different ingredients used in the manufacturing process. Some people who become sensitized to penicillins or related compounds at work also react to penicillin antibiotics taken by mouth. This implies that infections in such workers are best treated by antibiotics which are not related to the penicillins.

Cotton and other vegetable fibres

Although textiles have been made from cotton and flax since times of antiquity, respiratory symptoms due to exposure to cotton dust (byssinosis), were not recorded until the introduction of mechanized manufacturing processes in the early nineteenth century. Byssinosis can arise from exposure to a variety of vegetable fibres including:

● cotton

● soft hemp

● flax

However, the disease is not caused by the vegetable fibres themselves but by impurities.

Cotton Inhalation of dust causing byssinosis occurs mainly in:

(1) the ginnery—where a machine, the 'gin', removes seeds from the cotton;

(2) the mixing room—where bales of cotton are opened and mixed;

(3) the card room—where machines (carding engines) comb the fibres to produce the initial thread, but at the same time release impurities.

Dust in the air of cotton mills, nearly all of which in the UK are in Lancashire, consists of broken cotton fibres, fragments of the cotton plant, bacteria, and fungi. 'Coarse' grade cotton produces more dust than 'fine' grade. Although the exact nature of the component in cotton dust that causes byssinosis is not known, it is likely that toxic materials (endotoxins) from bacteria are likely to play an important role.

Flax Flax is used less extensively since the advent of synthetic fibres, but it is still used to manufacture linen (an old English word for flax), twine, thread, and rope, particularly in Northern Ireland. In modern processing, small bundles of flax are separated out from bales by hand, mixed, and passed for automated 'hackling' where fibres are combed and impurities excluded. Following this, the fibres are carded, straightened on a drawing frame, twisted, and wound onto bobbins. Winding, twisting, and cabling of rope generates much less dust than opening, hackling, carding, and spinning.

Hemp   Only soft hemp (*Cannabis sativa*), also known as English or Irish hemp, is associated with symptoms of byssinosis. It is used in the production of rope and yarn. Unlike flax, hemp is batted (beaten) to remove wood particles before it is hackled and baled. Batting and hackling are very dusty activities as opposed to cabling, twisting, and polishing, where dust levels are low.

Drugs

Adverse reactions to drugs may be due to a direct toxic effect, intolerance, or allergy. Indeed only a small proportion (5 per cent) of drug reactions are due to allergy. Since any drug can be involved it is impossible to compile comprehensive lists of drugs implicated in allergic reactions. In the nineteenth century there were few effective drugs (e.g. digitalis, quinine, and morphine) and these did not consistently produce allergic reactions. Many new synthetic drugs were introduced just as the field of allergy was developing and reactions to them were more accurately documented. Drug allergy was not, however, an important problem until the highly allergenic sulphonamides were introduced in the 1930s. On average, 2–3 per cent of courses of drug treatment are complicated by allergic or allergy-like reactions.

True allergic reactions occur when large protein molecules induce antibody synthesis in the body. The majority of drugs are much smaller molecules, and the drug, or breakdown product, must combine with a protein already present in the body before it is capable of causing antibody production and an allergic response when the drug is encountered again. This is the most likely way in which drugs are involved in true allergic reactions. Often patients notice reactions on first exposure to a drug which cannot be due to allergy unless the drug had been previously encountered in different circumstances, for example ingestion of penicillin in milk.

Drugs which are most commonly associated with allergic reactions include:

● cephalosporins
● penicillins

Often drug reactions thought to be 'allergic' do not have the immunological basis described above. Indeed, some drugs are capable of producing typical 'allergic' symptoms without the production of antibodies—pseudo-allergy—by acting directly on mast cells to produce chemical mediators, e.g. histamine. Chemicals which may act in this way to mimic an allergic response include:

(1) iodinated contrast, media used in some X-ray examinations;

(2) muscle relaxants, particularly suxamethonium.

Aspirin intolerance
Aspirin (acetyl salicylic acid) is one commonly used and important drug to which people may be intolerant, and it is featured in one of the earliest publications on drug allergies, produced in 1919. Ironically, today there appears to be little evidence that aspirin intolerance involves a strictly 'allergic', or immunological, response. The first occurrence of a reaction to aspirin was reported in 1902 shortly after its introduction. The frequency of reaction varies according to the patient population sampled and the symptoms considered to represent intolerance. Approximate incidences are:

patients with rhinitis only	0.7%
patients with asthma only	4.3%
all patients with asthma and/or rhinitis	2.4%

Curiously, the incidence in females is slightly greater than that in males. This may be because it is possible that salicylic acid is the problem compound causing reactions. Salicylic acid is released from aspirin (the parent compound) as a result of the action of aspirin esterase, an enzyme which occurs in higher levels in females and is therefore probably released more quickly and in greater quantities in females than in males.

4

WHAT MAKES PEOPLE ALLERGIC?

Allergic diseases such as asthma, rhinitis (nasal allergy), and eczema are important world-wide. Allergy is usually considered to be an inappropriate and harmful response of the immune system to normally harmless substances. There are several basic mechanisms by which allergic symptoms may occur, but Type I allergy, which depends on an interaction between an allergen and immunoglobulin E (IgE) antibody, is most frequently implicated in the common allergic diseases. An individual capable of reacting to ordinary exposure to allergens in this way is referred to as 'atopic'. This can readily be demonstrated by skin-testing with extracts of the common inhalant allergens. Atopy is a word derived from the Greek *a topos*, meaning 'no place', and was translated by Coca and Cooke in 1923 to mean 'strange disease'. The 'strange disease' in question was allergy which appeared to be hereditary, limited to humans, and expressed as asthma or hay fever. Examination of these early ideas about allergic disease can help to provide some answers to the vital question, 'What makes people allergic?'; or, to put it another way, 'Who gets allergy?' Originally, IgE or 'reagin' as it was first known, was thought to be unique to atopic people. Now it is known that not only is IgE a normal immunoglobulin in all humans, but that animals also produce IgE. However, allergic individuals have higher levels of IgE than other people. This leads to the question of why this should occur and what, in those who do have high levels of IgE, dictates whether they will develop asthma, rhinitis, or neither.

What part does inheritance play?

Evidence from family studies and twin studies indicates that allergic diseases are hereditary. The first serious scientific study of the epidemiology of allergic disease, published in 1916, showed that almost half (48 per cent) of people suffering from allergies had a family history of allergic problems, compared to only 14 per cent of

HEREDITY

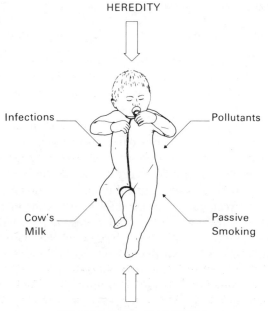

Infections

Pollutants

Cow's
Milk

Passive
Smoking

EXPOSURE TO ALLERGENS

Fig. 10 Factors affecting expression of allergy: 'what makes people allergic?'

non-allergic patients. Later studies have also shown that half to three-quarters of allergic patients have allergies 'in the family'. This suggests that inheritance has some part to play in the development of allergic disease. It is now known that individuals do not inherit a specific disease, such as asthma or rhinitis, but, rather, they inherit the tendency to develop high levels of IgE to common allergens. The way in which this is expressed as a disease, or whether disease develops at all, is determined by other factors. These may be additional inherited factors, or environmental influences. At present it is still not known which of these dictates whether a person with high levels of IgE against an allergen, such as grass pollen or house dust mite, will develop allergic disease, and it is more likely that both are important.

Couples who suffer from allergies are often concerned about the risk of their children becoming allergic. The risk appears to vary in different populations but approximate figures are:

neither parent allergic	10–20%
one allergic parent	30–50%
two allergic parents	40–75%

Where only one parent is affected it appears to make no difference which parent is allergic. It must be emphasized once more that it is the tendency to produce high levels of IgE against allergens which is inherited. There is no evidence that the materials to which a child develops allergy, and positive skin-tests, are influenced by the particular materials to which a parent is allergic. For example, the child of a parent who is primarily sensitive to grass pollen may experience most problems from cat dander or vice versa.

Studies of 7000 pairs of identical twins have shown low levels of agreement between each member of a set of twins in terms of the allergic diseases from which they suffer, despite their having exactly the same inheritance from their parents. If one twin has allergic rhinitis the other twin also has the disease in only one out of five cases. This confirms that the development of allergic disease must have an environmental as well as a hereditary component. More evidence for this is provided by the fact that differences in the prevalence of allergic disease in the community relate more to geographical location than racial origins. For instance, children of Afro-Caribbean origin born in the UK have the same prevalence of allergic asthma as European children born in the UK, whereas the prevalence is much lower in Afro-Caribbean children born in the West Indies.

Another interesting question concerns the likelihood of a patient who already suffers from one allergic disease developing additional symptoms. This risk is lower than many people expect. For example, a hay fever sufferer is only 2–3 times more likely to develop asthma than the general population. Indeed, only about 10 per cent of hay fever patients will ever develop asthma.

What part does environment play?

It is now clear that differences in the prevalence of allergic diseases are primarily related to environmental factors rather than to racial or other genetic factors. The nature of the different allergens encountered by an atopic individual has a profound effect on the development of allergic disease. In the USA, hay fever develops for the first time in 16 per cent of immigrant Chinese students during their time at university. In contrast, most American students who are predisposed to allergies have already developed hay fever by the age at which they enter university, and new cases of hay fever occur in only 1–2 per cent of students. This is due to the fact that immigrant students are meeting American allergens, such as rag-weed pollen, for the first time and developing a sensitivity soon after this first encounter.

A similar phenomenon has been recorded in Kuwait, where introduction of the Mesquite tree, which is not an indigenous species, resulted in an increase of respiratory allergy, following exposure to this new allergen, in people who previously had been well. This effect is also seen in developing countries such as Papua New Guinea where the prevalence of asthma in the adult population has risen almost seventy-fold in the past 5–10 years. All the asthmatic patients have positive skin-tests to *Dermatophagoides* mites. This being so, it appears that the change in lifestyle most likely to have influenced the increase in allergic asthma is the introduction of cotton blankets, which are used completely to encase the person during sleep, are rarely washed, and harbour very high numbers of mites.

Allergens indigenous to a person's normal environment are obviously encountered at a very early age. To investigate the influence of these early allergen contacts, the development of atopy in new-born infants from families with a history of allergic disease has been studied during the first year of life. The number of children with positive skin-tests to common allergens was found to rise with age:

- 16 per cent at 3 months,
- 18 per cent at 6 months,
- 33 per cent at 12 months.

In total, almost half (48 per cent) of infants showed positive skin-tests at some time during their first year, most of them to more than one allergen. Some children who were positive at 3–6 months were negative by one year, and it may be that these were children 'growing out' of their allergy very early in life. Contact with allergens in early infancy is obviously important in the development of allergic disease. For example, the keeping of pets in the home results in 10 per cent of infants of allergic parents in Britain becoming sensitized to cat dander before they are one year old.

It has been suggested that, since early allergen exposure is so important, there may be a relationship between month of birth and the subsequent development of allergic disease, since many allergens show seasonal variations in level. Studies in Finland have shown that this may hold true, with sensitization to birch pollen associated with birth in the months February to April when pollen levels are at their highest. Surprisingly, this effect was only seen in males (it is difficult to explain why this should be). Nevertheless, it has been calculated that if everyone aged 0–19 years had been born in the months least associated with pollen allergy (July–August), birch pollen allergy in young Finns could be reduced by 25 per cent.

Investigations in the UK have produced less clear-cut results with respect to the 'month of birth effect', some studies showing that skin reactions to *any* allergen were more common in people born between September and October than in those born in the rest of the year. This raises the question of whether the month of birth effect depends on some seasonal factor other than early exposure to particular allergens. For example, the risk of developing allergy may be greater for those who pass through the winter season of respiratory viruses in early infancy (see next page). Selection of patients for inclusion in 'month of birth studies' has always been a problem, some investigators including everyone with positive skin-tests to an allergen, and others including only those patients suffering actual symptoms. This has almost certainly confused the issue, and possibly obscured any true associations with month of birth in allergic patients. Further evidence in support of a real association comes again from Finland where allergy to birch pollen has been shown to correlate with severity of the first pollen season to which an infant is exposed.

The first few months of life, when the body's immune system is still immature, appear to represent a susceptible period when environmental influences have a profound effect on the subsequent development of allergic disease. The mechanism by which this could occur is, however, not fully understood. It is thought that the development of IgE in atopics occurs following absorption of small amounts of allergens through the mucous membranes lining the respiratory and gastrointestinal tract. At one time it was thought that membranes were more permeable, or leaky, in atopic people, allowing easier access of allergens into the body and therefore increasing the chance of production of IgE antibody. This is no longer thought to be the case but, rather, that atopic individuals are simply good producers of immunoglobulins in general, including IgE. Nevertheless, certain factors, such as viral or bacterial infections and cigarette smoking, can influence the permeability of the respiratory tract and are associated with the development of increased IgE production and, possibly, allergic disease. Whether this is due to changes in permeability or to other effects on the immune system is not known.

Studies in the late 1970s have investigated the influence of respiratory tract infections on the development of allergic symptoms in 13 children born into families where both parents were allergic. Not surprisingly, all but two of the children developed allergic symptoms within the first year of life. In nine out of 11 cases the allergic symptoms first appeared within six weeks of a respiratory tract infection (cold or 'flu), suggesting a possible relationship between such infections and the onset of allergic disease.

Virus infections cause extensive damage to the respiratory mucosa, the layers of cells which line the respiratory tract. This being so, it is possible that this damage allows inhaled allergens to pass more easily into the tissues where they can make contact with antibody-forming cells. At present there is no definite proof of a causal relationship between viral infections and the development of allergic symptoms. It is still possible that there is instead some common factor which predisposes infants to both frequent infections and allergic disease. If viral infections do act as a precipitating factor for the development of allergic symptoms in predisposed individuals, the exact mechanism of their action is still unknown. However, support for the idea that their effects result from mucosal damage comes from observations of the development of occupationally related allergic disease.

In adult life, exposure to new allergens is less frequent than in infancy but does occur in the workplace when individuals come into contact with materials not encountered in daily life. In this situation another agent which causes damage to the lining of the respiratory tract has been shown to play a part. Cigarette smoking has been observed to increase the likelihood that atopic people employed in the electronics industry will become sensitized following exposure to phthalic anhydride. It is possible that this results from smoking causing mucosal damage similar to that caused by viruses in infancy, again allowing allergens to come more easily into contact with antibody-forming cells.

At present, more work is needed to investigate the influence of factors such as viruses, tobacco smoking, and atmospheric pollution on the development of allergic disease in predisposed individuals.

Does breast-feeding prevent allergy?

It is well known that breast-feeding strengthens bonding between mother and child and it is strongly recommended by paediatricians for this reason alone. In addition breast milk contains various antiseptics, including a special antibody called immunoglobulin A, which prevent bacteria and viruses from multiplying in the baby's digestive system and causing diarrhoea. Non breast-fed babies are also at greater risk from colds and bronchitis. For this reason it is important that babies, particularly those born in the Third World, continue to be breast fed.

The energy we all need for our growth and existence comes entirely from foreign fats, carbohydrates, and proteins in our diet. Proteins especially are highly allergenic, and it is remarkable that we can tolerate so many without severe allergic reactions. On the face of it, one mouthful of hen's egg should be more than enough to cause severe wheezing, skin rashes, and a fall in the blood pressure— a collection of symptoms called anaphylaxis—which can be fatal. The fact that this does not happen, except to an unfortunate few with severe food allergy, is due in part to the breaking down of food allergens into their constituent chemicals which are too small to be allergenic. The lining of the normal human digestive system largely prevents anything but these small chemicals from being absorbed.

Quite apart from the unpleasantness and dangers of diarrhoea in young babies, it is now known that such infections make the baby's gut more permeable, allowing undigested food particles to be absorbed causing allergy. In addition the baby's digestive system, being immature, is more permeable or leaky, and will allow more foreign allergens to enter the blood stream than an adult's would. Under these circumstances it is obviously sensible to restrict the baby's diet to the least allergic food, mother's milk.

Many people believe that the introduction of artificial feeds in the form of cow's milk has had a detrimental effect on the prevalence of allergic disease. Eczema is common in babies, particularly aged between two and six months, and asthma frequently occurs in the first two years of life, when it is both difficult to diagnose and treat. The main question is whether exclusive breast-feeding for the first three to six months of a baby's life can influence the development of eczema and asthma. In general the answer is 'yes', particularly for mothers who themselves suffer from allergic diseases or come from a family where these diseases are present. Some indications of the likely development of allergic disease in a baby can be obtained by testing blood from the umbilical cord for IgE—the antibodies involved in allergy. If the levels are high then the chance of the baby developing eczema in the next few months is increased. Studies from round the world have indicated that exclusive breast-feeding should be continued for at least three months and, if possible, six months. This regime is very difficult for most mothers to sustain but, unfortunately, even a single supplementary feed with infant foods based on cow's milk can initiate allergy. This problem can be overcome by feeding on occasions with infant formulas containing soya milk.

Although exclusive breast-feeding is beneficial in reducing the incidence of allergic diseases, there is growing evidence that proteins from food eaten by the mother can reach the baby, across the placenta during pregnancy and through breast milk, in sufficient quantities to initiate allergy. It has been shown that complete avoidance of highly allergenic foods, such as milk, dairy produce, egg, fish, beef, and peanuts, throughout pregnancy enhances the beneficial effect of breast-feeding in preventing infantile eczema.

Although dietary restrictions during pregnancy and lactation may be beneficial in reducing the risk of the baby developing allergic disease, such a restricted diet should not be undertaken without medical supervision to ensure that the baby does not lack any chemicals needed for normal, healthy growth and development.

Is the allergic state in any way beneficial?

Many people believe that the atopic state, or allergic disease, is becoming increasingly common in the population. It is possible that this is due to a greater recognition of these diseases by both patients and the medical profession, rather than a true increase in prevalence. It does seem, however, that levels of atopy in the population are unlikely to be declining. This situation has prompted debate as to whether allergy has a biological role in the population, since, as Darwin explained, '. . . every slight modification, which . . . chanced to arise, and which in any way favoured . . . individuals . . . , would tend to be preserved.'

It is difficult to see an advantage in bouts of sneezing or wheezing in response to grass pollen. However, in some circumstances an enhanced or more rapid recognition of foreign proteins in the body, such as parasitic worms, could be advantageous. In primitive societies it seems unlikely that symptoms caused by common inhaled allergens would be too high a price to pay for the advantages of a rapid defence against worm infestations. It is an attractive idea that, in more civilized societies, increased levels of local immunity in atopic people confer some other advantage, since worm infestation is relatively uncommon. To investigate this we need to ask the question 'Are some diseases less common in atopics than in the general population?' Some people think that the atopic state reduces the risk of developing malignant (cancerous) tumours. Circumstantial evidence in support of this idea comes from the fact that the incidence of allergy decreases with age whereas incidence of cancer tends to increase. However, this does not confirm a link between the two diseases. More objectively, skin-tests with extracts of an individual's own tumour (removed at surgery) have shown positive reactions, indicating that it is possible for the atopic patient to become sensitized to the 'foreign tissue' of the tumour. Scientific studies of the potential protective effect of allergy have obtained conflicting results, some studies reporting no association between allergy and cancer, and some showing a lower incidence of allergy in cancer patients.

If the atopic state is associated with any benefit to people in the developed world, the nature of this advantage has yet to be demonstrated.

Part II

HOW CAN ALLERGY AFFECT THE BODY?

5

HAY FEVER

The nose and eyes are the parts of the body most frequently affected by allergy and there is evidence that the number of people suffering from allergic complaints involving these organs is increasing. In a recent survey of over 300 000 patients in general practice, the number of people visiting their family doctors for treatment of hay fever was found to have doubled in the past 10 years.

Of all medical terms, 'hay fever' is the best known to those outside the health professions, but unfortunately it does not accurately describe the condition to which it refers. The problem regarding the definition of this disease is to distinguish it from other causes of 'catarrh', such as the common cold. The symptoms of hay fever and the common cold are virtually identical. Sufferers from both have to endure the miseries of copious watery discharge from the nose, sneezing, nasal blockage, sore, runny eyes, and often pains in the face and headache due to congested sinuses.

Hay fever has been recognized since the sixteenth century, but it was not until the early nineteenth century that the classical features of the disease were first described by Dr John Bostock from Guy's Hospital, who was himself a hay fever sufferer. He recognized the seasonal nature of hay fever and described it as *'catarrhus aestivus'*—'summer catarrh'. There were many suggestions at the time as to the cause of this mysterious disease, which included 'odoriferous emanations' from grasses and hay. By 1835 the term 'hay fever' had found its way into the English language, but it was not until 1873 that Charles Harrison Blackley, using himself as a guinea-pig, was able to show that hay fever symptoms were due to the effects of pollen grains on the nose and eyes.

Charles Blackley was born in Lancashire in 1820 and trained as an engraver, eventually starting his own business. Following his marriage in 1855, his interest in science prompted him to change profession and study medicine, qualifying in 1858 and setting up practice in Hulme, near Manchester. Having proved that he could reproduce his own symptoms by putting pollen grains in his nose and eyes, Blackley became interested in methods of collecting grains on glass microscope slides covered with glycerine, which caused the grains to stick, and carbolic, which discouraged insects who were taking too keen an interest in his work.

One of the things which intrigued Charles Blackley most was the way that he would often get symptoms even when the wind was blowing consistently from a direction where he knew that there was no grass for several miles. He began to wonder if pollen could be transported high in the air over large distances and tested this idea by attaching his slides to kites which he flew at different altitudes. By flying two kites 'in tandem' he achieved a height of 2000 feet! He found that pollen counts at 1–2000 feet were higher than at ground level, confirming his suspicion that airborne pollen could be carried over great distances.

Charles Blackley was not the first person to suspect the relationship between pollen grains and hay fever, but he was the first investigator to demonstrate it experimentally. Nowadays this causal relationship is well recognized and the word 'hay fever' is less frequently used among the medical profession, who prefer a name which more accurately describes the seasonal nature of the disease and its cause. Probably the best term for the disease is 'seasonal allergic rhinitis (and conjunctivitis)', which can be further qualified by addition of the appropriate phrase 'due to grass pollen' or 'due to tree pollen'.

Nasal symptoms

Nasal symptoms which persist throughout the year represent a disease which is distinct from 'hay fever' and is best described as 'perennial rhinitis'. The causes of perennial rhinitis can be various, including animal and occupational allergens, but the most frequent cause is the house dust mite.

In some patients with nasal symptoms, no allergic cause can be found. They are often said to be suffering from vaso-motor rhinitis. In these sufferers, like those with nasal allergy, the nose is more sensitive to a wide variety of stimuli, such as cold air, cigarette smoke, traffic fumes, strong smells from disinfectants, perfumes, aftershave, and paints, and many types of fine powders and dust. A stuffy nose can also result from drinking alcoholic drinks and sexual arousal, both of which increase blood flow to the erectile tissue in the nose.

Itching

Itching or tickling in the nose leads to bouts of 5–20 sneezes. Itching of the roof of the mouth, or soft palate, results from pollen swept out of the nose into the back of the throat. Some people also suffer from itchy ears. This is not due to pollen landing in the ear but is the result of activation of a common nerve that links the back of the throat and ear. Itching and sneezing result from the effects of histamine released in the nose during the allergic reaction to pollen. The most important role of the nose is to filter air and protect the delicate tissues of the lungs from pollutants, and sneezing not only expels large particles from the nose but is a warning of the presence of harmful irritating material in the atmosphere. Irritation of just a very small area of the nose can result in instantaneous and vigorous sneezing. Unfortunately, this sensitive and protective mechanism is triggered in patients with hay fever at the height of the season by the inhalation of just a few grains of pollen.

Runny nose

The nostrils are kept clean by secretion of a sticky fluid. The human nose can secrete up to an egg-cupful of fluid every hour, which spills out, necessitating constant sniffing and nose blowing. Like sneezing, this mechanism is also caused by histamine and can begin rapidly after pollen is breathed into the nose. Inhalation of pollen throughout the day leads to a constant runny nose. All of us have experienced the effects of excessive nasal secretion, usually noticed as mucus in the back of the throat which is normally swallowed. Some people prefer to spit this out into a handkerchief and think that it is phlegm from the chest.

Stuffy nose

Throughout the day the width of the nostrils varies in a cyclical way, so that one is always more open than the other. These changes, occurring every two to four hours, result from alteration in the amount of blood flowing through the lining of the nose and allows one nostril to rest. This can readily be tested by sniffing first through one nostril and then the other. We are normally only aware of the effects of this cyclical change in the patency of the nostrils during colds or attacks of hay fever when a nostril, which is already partially closed, becomes totally blocked. This explains why colds appear to move from one nostril to the other. The cause of a stuffy nose during colds and hay fever differs to some extent, in that the accumulation of nasal secretions is more important in colds than in hay fever, and can therefore be relieved by nose blowing. In hay fever, however, nasal blockage is mainly due to changes in blood flow in the lining of the nose, leading to swelling of the tissues, which is little helped by blowing the nose. A completely blocked nose causes headache, disturbed sleep, mouth breathing, and a dry throat and tongue on waking. In severe cases, sufferers cannot smell or taste. These disruptive symptoms are more common in those who suffer from perennial rhinitis rather than hay fever.

All of us are aware of the obvious symptoms, such as sneezing, which accompany hay fever, but some aspects of rhinitis are less well known. For instance, many children with long-standing allergic rhinitis show distinctive facial characteristics and mannerisms. They often have 'bags' under their eyes. Many children wipe their noses with the palms of their hands—the allergic salute—which eventually can lead to a crease across the nose.

Sinusitis

A persistent blocked nose often leads to problems with the sinuses. These are air-filled cavities which lighten the skull, helping us to hold our heads up and adding resonance to our voices. The lining of the sinuses is the same as that of the nose, and produces secretions which drain through small holes into each nostril. These openings easily become blocked during colds, hay fever, or perennial rhinitis, causing headaches and pain in the face. Sometimes just one of the sinuses can be blocked, causing symptoms only on one side of the head. Bacteria trapped in the blocked sinus multiply and cause infection—acute sinusitus, which is not only very painful but causes a fever and makes people feel generally off-colour. Sinus trouble always results from nasal disease and clears up when the nose problem is treated.

Allergy and the eye

Everyone is familiar with the severe running and itching of the eyes

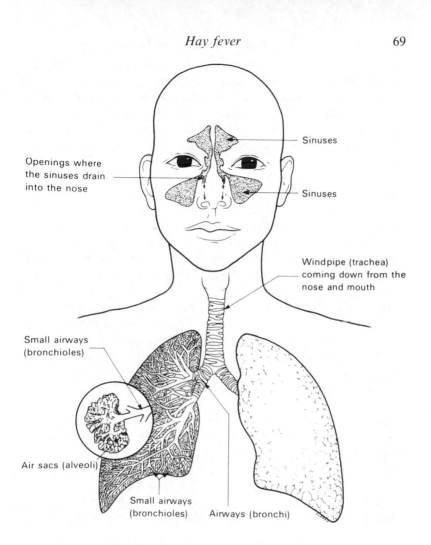

Fig. 11 Sinuses and respiratory tract.

that accompany hay fever, but in fact the eye is unique in that most of it cannot mount an allergic response. Any inflammation in the eye itself would cause loss of vision, and so there are few blood vessels or mast cells present. In certain circumstances this is helpful, for example to those patients who have injured the transparent region in front of the lens, the cornea, which can easily be replaced by a transplant with no fear of rejection. Obviously, the inside of the eye must be protected since it cannot adequately protect itself through the usual defence mechanisms. To this end, the cells of the cornea are stuck tightly together, preventing particles, including allergens and bacteria, from entering. In addition, the front of the eye is continuously wiped clean by secretions (tears) and blinking. The secretions contain powerful disinfectants, one of which is called lysozyme, and antibodies, particularly IgA, which react with and neutralize allergens and bacteria.

However, the conjunctiva, the membrane which covers the white part of the eye and lines the inside of the eyelids, can and does mount a vigorous allergic response to external agents. Although there are no mast cells in the cornea, many are present in the conjunctiva. It readily becomes inflamed, the blood supply increases, and the eye looks red. More tears are secreted to wash away the offending substance, causing running of the eyes. Particles accumulate in the inner corner of the eye and are removed by rubbing, an act encouraged by itching. Although these mechanisms are vital for protection of the eye against infection, their persistence due to allergens present in the air leads to the troublesome symptoms of allergic conjunctivitis.

Allergic conjunctivitis

The commonest cause of this complaint is grass pollen, and the worst symptom is itching, particularly of the inner corners of the eye where the allergen particles have been swept. This leads to persistent eye-rubbing, which itself causes further reddening of the eyes, and swelling and damage of the skin of the eyelids. Unfortunately, rubbing only relieves itching for a short while and is best avoided, but this is difficult.

As with the nose, irritation of the eye can arise from non-allergic stimuli, such as smoky atmospheres. In this case, the irritation usually results in smarting rather than the intense itching which is characteristic of allergic symptoms. Allergy does not affect the transparent cornea which covers the iris and pupil, and therefore intolerance of strong light—photophobia—is not usually a problem but can occur transiently after the eyes have been rubbed particularly vigorously.

Watery discharge from the eyes can result from irritation of the conjunctiva itself or from irritation of the lining of the nose (this is one of the reasons why peeling strong-smelling onions can result in floods of tears—something that can be avoided by breathing through the mouth rather than the nose). Eye involvement is more common in hay fever than in perennial rhinitis due, for example, to allergy to the house dust mite. Nevertheless, it can occur, and may be troublesome in children, as a consequence of allergy to pets.

Vernal conjunctivitis

This is a more serious, but fortunately rarer, form of conjunctivitis, though its name is a little misleading since vernal means spring and yet this condition continues throughout the year. However, the disease often flares up in the spring-time causing severe symptoms. Patients are usually atopic young boys less than 10 years old living in the warmer parts of the world. Allergy accounts for the flare-ups but not for the whole disease, the underlying cause of which is not yet understood. The main symptoms are persistent itching, thick, ropy discharge, and intolerance of light, since in this prolonged disease the cornea finally becomes affected. For the doctor, the most noticeable feature of the disease is the characteristic cobblestone appearance on the inner surface of the upper eyelid, due to swelling of the tissues, which form fleshy protuberances. Fortunately, this condition is self-limiting, resolving after 5–10 years.

6

ASTHMA AND LUNG DISEASES

Every cell in our bodies needs oxygen. In combination with sugar, oxygen releases energy, with the production of carbon dioxide as the waste product. This process is called 'aerobic respiration'. Simple organisms containing just one cell, such as an amoeba, can extract oxygen directly from the water in which they live. Similarly, carbon dioxide leaks out into the water through the enclosing cell membrane. Respiration in more complex organisms, such as man, requires special systems for delivery of oxygen from the air and removal of carbon dioxide. This involves the respiratory system, which includes the nose and lungs, and the cardiovascular system, which consists of the heart and blood vessels. The constituent parts of the respiratory system are shown in Fig. 11. The purpose of the nose is to filter dust, bacteria, viruses, and chemicals from the air, and to warm and moisten the air before it reaches the more delicate tissues of the lung. Air is sucked through the nose and into the lungs by the co-ordinated movements of the muscles around the ribs, which pull the ribs upwards and outwards. Tightening of the muscles in the diaphragm pulls the diaphragm down, increasing the volume of the chest. This combined activity causes air to rush through the nose into the lungs. Relaxation of the muscles has the reverse effect. It is possible to exhale forcefully by contracting the muscles in the stomach wall and around the ribs, as in blowing out a candle or coughing.

Air on its way in and out of the lungs passes between the vocal cords in the larynx, where alterations of the waveform help to produce speech. Air moves down the windpipe (trachea), which divides into the right and left bronchi just below the breastbone. The bronchi (which supply the lungs), like the trachea, have firm, though incomplete, supports made of cartilage, which hold them open. Each bronchus then divides a total of 25 times before reaching the critical part of the lung, called the alveolus, where exchange of oxygen and carbon dioxide occurs. After the first 10 divisions, the smaller

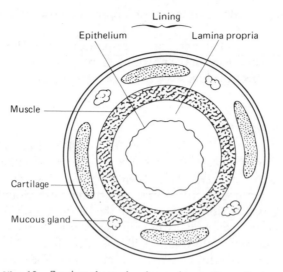

Fig. 12 Section through a lung airway (bronchus).

bronchi have no firm supports and are easily compressed as we exhale and the pressure inside the chest increases. Quite why these vital air passages lack support in man is a mystery: it certainly confers no benefit. In fact it is the reason why breathlessness occurs in diseases such as asthma, chronic bronchitis, and emphysema, where damage causes them to narrow and compress more easily. Although these smaller airways have no cartilage to hold them open, they do, in common with the larger airways, have an encircling layer of muscle, as shown in Fig. 12. This gets progressively thinner as the airways get smaller and, although it gives some firmness to the tubes, contraction of this muscle is part of the cause of asthma. Once air reaches the six hundred million alveoli in the lung, it is spread extremely thinly over an area the size of a tennis court, allowing plenty of room for gas exchange to occur.

Overall the narrowest part of the respiratory tract is the nose. This is obvious during exercise when breathing always occurs through the mouth to allow sufficient air to be rapidly moved in and out of the lungs. If it were not for the mouth, hay fever would be a fatal disease! The next narrowest part is in the larynx between the vocal cords which, as mentioned on page 105, can be affected by allergy causing angio-oedema. It is easy to appreciate how serious this can be.

Allergy can affect all parts of the respiratory tract. Hay fever has already been discussed in the previous chapter. Now we will see how allergy can affect the lungs.

Asthma

Asthma through the ages

The word 'asthma' is derived from the Greek and literally means 'hard breathing'. The first detailed accounts of asthma occur early in the Christian era when Galen, the most famous physician of the second century, described an intermittent obstruction to breathing caused by secretion dripping from the brain into the lungs. This view about the cause of asthma persisted until the seventeenth century, and many physicians still consider that asthma can be caused by secretions dripping from the nose, if not the brain, into the lungs. The foremost physician of the twelfth century, called Moses Maimonides, wrote in his 'Treatise on the Asthma' that the disease frequently began at puberty. Although he was right in this respect, he also thought that the most effective remedy was soup made from fat hens! Our commonly held view about the importance of the environment as a cause of asthma, was first recognized by an Italian medical astrologer called Girolamo Cardano, who, in 1552, cured an archbishop's asthma by removing his feather bed and feather pillows.

The first book devoted entirely to asthma was written in London in 1698 by Sir John Floyer, and his account of the difficulties in both diagnosing and treating the disease are almost as true today as they were three centuries ago. More recently, in the nineteenth century, Dr Salter noted that exercise, cold air, dust, pungent fumes, and 'animal and vegetable emanations' were important causes of asthma. Other physicians had different ideas, particularly Dr William Osler, one of the greatest physicians of the early twentieth century. He wrote that 'all agree that there is, in a majority of cases of bronchial asthma, a strong neurotic element' and by so doing placed an unfortunate stigma on this disease which still permeates much current thinking.

Modern views on asthma

The *Oxford English Dictionary*'s definition of asthma includes:

- intermittent difficulty with breathing;
- wheezing;
- constriction in the chest;
- cough and expectoration.

However, asthma is not the only cause of wheezing, and a productive cough is more commonly a symptom of chronic bronchitis. It is important to diagnose asthma properly since there are many different causes of attacks of breathlessness, cough, and wheeze, each requiring a different treatment:

(1) heart disease—best treated with diuretics;

(2) acute bronchitis—best treated by antibiotics;

(3) chronic bronchitis—best treated with antibiotics and advice on stopping smoking;

(4) farmer's or bird fancier's lung—best treated by removal of the cause of the problem;

(5) asthma.

Basically, asthma refers to a condition in which there are rapid changes in the diameter of the airways of the lung, characterized by one or more of the following symptoms:

- wheeze
- shortness of breath
- constriction in the chest
- cough (sometimes productive)
- night-time symptoms

accompanied by reversible changes in airway diameter shown:

(1) spontaneously by measuring breathing throughout the day, as described on page 78;

(2) after treatment with bronchodilators or corticosteroids, as described in Chapter 11;

(3) after exercise or inhaling cold air, as described in Chapter 9;

(4) after special inhalation tests in the hospital, also described in Chapter 9.

Symptoms

Shortness of breath   This is one of the main symptoms of asthma, but the degree of breathlessness experienced depends on:

(1)   Lifestyle. Obviously the mildest asthma in an athlete can seriously affect performance and will be noticed straightaway. In contrast, asthma in sedentary workers who drive to work and spend evenings watching TV will not be noticed until the disease is more severe. Nevertheless, many people notice their asthma for the first time when exercising, for example when walking home from work or carrying the shopping.

2.   Perception. People vary greatly in their perception of sensations arising from the lungs. While some are able to continue at work or exercise in spite of severe asthma, others feel quite disabled by the mildest disease. This is illustrated in the following case.

A 40-year-old bank employee, who refereed one or two football matches every weekend and considered that he was quite fit. His bank introduced a system of health screening for the over-40s, including measurement of lung function which, in his case, was less than half that expected for someone of his age and build. He was shocked to get this result, especially since he had never smoked. After asthma had been diagnosed he had to be persuaded to take regular treatment, since he did not consider that he had any serious lung disease. To his surprise the treatment transformed his life, making him much more alert, efficient at work, and able to referee his football matches without the tightness in the chest and breathlessness which he had previously considered to be normal.

Asthma occurring in schoolchildren can be particularly disruptive. Attacks of breathlessness prevent participation in games, and many asthmatics are put off exercise for life. Thanks to modern treatment, this need no longer be the case, and many well-known sportsmen and women are asthma sufferers.

Tightness in the chest   Many asthmatics experience odd sensations in the chest which they find hard to describe. Often there is a slight sense of choking or of difficulty in taking a deep breath. Some describe a heavy oppressive feeling, like a band around their chest. This symptom is similar to that described by patients who have heart disease giving rise to angina.

Wheeze   Wheezing is an extremely common symptom experienced by up to a quarter of the population at some stage in their lives. Wheezing is neither an obligatory nor diagnostic feature of asthma, though it does imply narrowing of the airways of the lung, and the commonest cause is a chest infection. Typically, the wheezing sounds are of many different pitches and are generated by air vibrating as it is blown out through narrowed tubes of different sizes. In very severe asthma the rate at which air is blown out of the lungs is so sluggish that wheezing does not occur.

Cough Until the Clean Air Act of 1953 Britain was a country plagued by air pollution. The combination of fog and sulphur dioxide from burnt fossil fuels gave rise to thick smogs which caused many people to die from chronic bronchitis. Indeed, this was known as the 'English disease'. Although the most important factor causing chronic bronchitis was, and still is, cigarette smoking, atmospheric pollution had an important role. With the decline in cigarette smoking and air pollution it has now been realized that asthma can cause a productive cough, and this is the most likely disease in a non-smoking patient with such symptoms. Only recently has coughing been recognized as one of the most important and early symptoms of asthma, particularly in children. All too often recurrent coughs in childhood are attributed to chest infections or bronchitis and treated with antibiotics. Although it is perfectly true that chest infections are common in infancy and childhood, prolonged illness lasting more than a week or occurring more than three times a year requires careful investigation. If recurrent infections are found to be the cause, then it is important to exclude serious underlying diseases like cystic fibrosis, which has a frequency of one in every 2000 children. Very rarely a deficiency of immunoglobulins, particularly immuno-globulin A, described in Appendix A, can be the cause. However, it is much more likely that the recurrent episodes of coughing will be due to asthma, which will respond rapidly to the correct asthma treatment, which is not antibiotics. At present it can take up to seven visits to the family doctor before the correct diagnosis of asthma is made. This is just what happened to James.

James, aged four, was taken to the doctor by his mother who was very worried by the fact that her son had been waking at night with a barking cough. When examined by the doctor James seemed well and not wheezy. He was prescribed a course of antibiotic syrup. In spite of this treatment, symptoms continued and the doctor prescribed another course of anti-biotics. No one else in the family had been suffering from a cold, and the boy's continuing night-time symptoms were causing considerable distress to his mother who was also sleeping badly. A cough linctus and steam inhalations were then suggested, and a chest X-ray taken by the local hospital was normal. At the next visit the doctor thought he could hear a wheeze and measured James's peak flow, which was normal. However, the doctor asked James to run several times around the surgery car park and subsequent measurements showed a large fall in his peak flow rate, indicating that the cause of his night-time symptoms was asthma. Regular treatment with Intal caused the symptoms to resolve completely.

Night-time symptoms   A peculiarity of asthma is that it is particularly bad at night. While some patients notice that they are worst on going to bed, others wake in the early hours of the morning complaining of a cough, shortness of breath, and a wheeze. Just why this happens is not fully understood, but it is related to the levels in the blood of a hormone called adrenalin. Adrenalin prepares the body for activity and its levels are high during the day but low at night. In addition, it relaxes the muscles around the airways of the lung, increasing their diameter and making breathing easier. The reverse is true at night when, in all of us, the bronchi are slightly narrower. This does not matter if our lungs are normal but becomes crucial when the airways are damaged and the muscles are already partially contracted as in asthma.

How asthma is diagnosed
It is vital to diagnose asthma properly and this cannot be done on symptoms alone. Tests are essential to demonstrate the reversible changes in airway diameter.

Peak flow records   Perhaps the best and easiest way to diagnose asthma is to blow into a peak flow meter, as shown in Fig. 13, several times throughout the day. The peak flow meter simply records the top speed at which air can be blown out of the lungs. The narrower the bronchi, the more difficult it is to force air out. This means that air leaves the lungs more slowly. Provided air is blown out of the lungs as hard as possible, the top speed is reached at the very start of expiration. This maximum speed is called the peak expiratory flow rate, or PEFR, and is measured as the number of litres of air which would be blown out if that speed could be maintained for a full minute. The peak flow rate also depends to some extent upon the strength of the muscles and diaphragm and particularly upon the original size of the tubes, so that the peak flow rate of an athletic young man will be almost twice that of a little old lady. The normal peak flow rates for men and women are shown in the charts on page

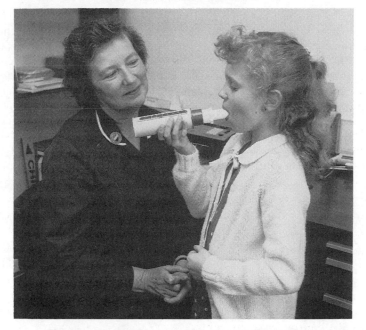

Fig. 13 Child using peak flow meter Mini-Wright.

128. In normal people the PEFR varies very little throughout the day, never falling by more than 15 per cent from the highest level, usually recorded in the middle of the day. In contrast, the PEFRs vary much more in asthmatic people. Once again the highest peak flow usually occurs in the early afternoon but PEFRs are especially low in the late evening and early morning. To diagnose asthma it is best to record peak flows three times a day: on waking, in the early afternoon, and before bed. Peak flow should be measured in this way for at least a week. Additional measurements should be made when symptoms are particularly bad, for example in the early hours of the morning.

The value of such measurements in making the diagnosis of asthma are illustrated by the following case history.

A 55-year-old usher at the Law Courts was referred to the hospital out-patient department because of a troublesome cough. He smoked one and a half ounces of tobacco a week in his pipe and had been diagnosed as suffering from chronic bronchitis by his own family doctor. He had found it difficult to stop smoking and frequent courses of antibiotics had not improved his cough, which was proving embarrassing in the Courts. His hobby was fishing and he denied any breathlessness. His lung function measured in the out-patient department was completely normal as was his chest X-ray. He was given a portable Mini-Wright peak flow meter to record PEFR in the mornings when he woke, in the afternoon at work, and before bed, for two weeks. When seen again, analysis of the results showed large swings in peak flow rate, making the diagnosis one of asthma. Treatment was begun immediately with inhaled corticosteroids in the form of Becotide, two puffs three times a day, which led to complete resolution of his symptoms and a considerable reduction in the diurnal variation of his PEFR. This man's PEFR chart is shown below.

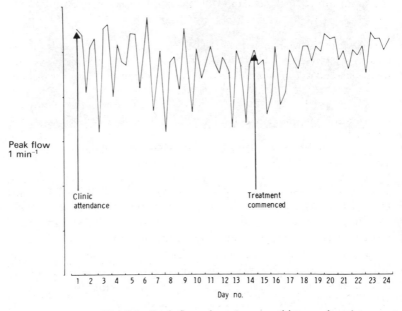

Fig. 14 Peak flow chart (see case history above).

In addition to making the diagnosis of asthma, PEFRs measured three times a day can be used to monitor the effects of treatment. Successful therapy will reduce the variability in the recordings, as shown in the case above. Apart from the variability in PEFRs, worsening asthma causes an overall decrease in peak flows. Such measurements can indicate when it is time to increase the treatment or visit the doctor. Most asthmatics find it very helpful to buy their own peak flow meter. Fortunately, portable meters such as the 'Mini-Wright' are compact, light, and inexpensive. They are available from Clement Clarke International Ltd, 15 Wigmore Street, London W1H 9LA.

Asthmatics rapidly become adept at using the meter and eventually can predict what their score will be from the way they feel. Nevertheless, a low peak flow confirms an attack and helps to convince sufferers that more treatment or a visit to their doctor is needed.

Regular daily recordings of PEFR can help identify causes of asthma. For example, peak flows may be low during the working week but improve at weekends if asthma is caused by sensitization to materials at work, as described in Chapter 3.

One 35-year-old woman had worked for the past five years in a factory making television sets. Her job was to solder printed circuit boards. She used a soldering iron and multicore solder. Walking home from work one day she noticed she was more breathless than usual, and she woke up in the early hours with a cough. Her symptoms worsened as the week went on but improved over the weekend. However, the following week her symptoms returned and did not respond to the antibiotics prescribed by her GP, who then referred her to hospital. She was given a portable peak flow meter and asked to record PEFRs three times a day for the next two weeks. Her chart showed clearly that PEFRs were much worse during the working week, and a diagnosis of occupational asthma was made. In this case the cause of the asthma was the flux, called colophony, included in the multicore solder, described on page 48. Fortunately, another job away from the soldering was found for her and her symptoms resolved.

Asthma is very hard to diagnose in babies and toddlers, who cannot blow satisfactorily into peak flow meters. In them the diagnosis usually has to be made on the story given by the mother, though sometimes wheezing can be heard in the chest. The other tests used to diagnose asthma are described in Chapter 9.

How common is asthma?

Asthma occurs in almost every country although it is extremely rare in American Indians and Eskimos. Over one-third of the inhabitants of Tristan da Cunha suffer from the disease. They are all descended from 15 ancestors, three of whom had asthma, indicating the importance of inheritance in this disease. In general, asthma is more common in developed countries compared to Third World countries, suggesting the importance of some environmental factors. Even in Europe its frequency differs from country to country. Part of this may be due to increased awareness and better facilities for diagnosis in certain countries, but there is little doubt that asthma is becoming more common. In the UK up to one in seven children is thought to suffer from asthma, which is commonest in boys. Asthma is less common in adults, although it still affects between one in 10 and one in 20 of the population and is just as common in women as in men.

What causes asthma?

It used to be thought that asthma had one particular cause, and so it was often labelled as being allergic, infective, or nervous asthma. We now know that although one particular cause may be the most important, attacks can be triggered by many different factors. Asthma primarily due to allergy to pets can nevertheless be brought on by a cold, exercise, a change in the weather, or emotional stress.

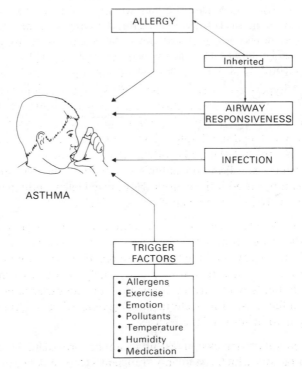

Fig. 15   Causes of asthma.

What all asthmatics have in common is that their airways are 'twitchy' and easily narrow when provoked by many stimuli which have no effect on the airways of normal people. For example, many asthmatics become wheezy on entering a room where people are smoking cigarettes. This characteristic of the airways of asthmatics is called increased, or 'hyper', responsiveness. It can easily be measured by the tests described in Chapter 9.

Airway responsiveness   Twitchyness of the airways occurs to a greater or lesser extent in all asthmatics. The more responsive the airways, the worse the asthma. In patients with rhinitis but no symptoms of asthma the airways in their lungs are often mildly hyper-responsive. This may only become apparent if they develop a chest infection which then makes them wheezy. Increased responsiveness of the airways is inherited, though not in a straightforward way. This indicates that environmental factors are also involved. One of these is allergy.

Allergens   The commonest allergen causing asthma in the UK is the house dust mite (this is fully described in Chapter 3, as are the other allergens that can cause this disease). Exposure of asthmatics to substances to which they are allergic makes their airways more responsive, whereas avoidance of these substances can reduce the degree of hyper-responsiveness, improve symptoms, and lessen the need for treatment. Certain dusts and fumes encountered at work can make people asthmatic. The important causes of occupational asthma are detailed in Chapter 3. The important points about such substances are that they, like other allergens, can increase hyper-responsiveness and bring on asthma for the first time. Occupational asthma can be cured if the cause is recognized early enough and the worker removed from exposure.

Infections   Viruses, such as those that cause the common cold, frequently bring on attacks of asthma. They cause damage to the lining of the airways making them more reactive. Antibiotics do not treat viruses and are therefore of little help in the treatment of asthma. Patients with chronic bronchitis, on the other hand, often develop infections with bacteria and become wheezy. Antibiotics certainly help in some of these cases.

Trigger factors   Attacks of asthma can be provoked by a wide range of factors which cause only transient symptoms because they do not alter the hyper-responsiveness of the airways:

● exercise

● emotion

● pollutants, perfumes, and aerosols

● air temperature and humidity

● drugs

Exercise   Most asthmatics wheeze after exercise, particularly if it is prolonged, for example jogging. The attack does not occur during the exercise but on stopping. Breathing cold, dry air will also cause an attack. In both cases the wheezing results from cooling and drying of the airways. On this basis it is obvious that swimming is the best exercise for asthmatics, unless the water is heavily chlorinated, when the fumes can be irritant.

Emotion   Everyone knows that stress can make asthma worse, but this does not mean that asthmatics have any more emotional problems than their non-asthmatic contemporaries. Relaxation can help to abort an attack but, unfortunately, a relaxed attitude to life rarely provides a complete cure.

Pollutants, perfumes, and aerosols   Many asthmatics know that a few breaths of cigarette smoke, car exhaust fumes, strong perfumes, or a dusty atmosphere can provoke an attack. Indeed, epidemics of asthma have occurred during periods of heavy pollution in industrial areas, caused by high concentrations of sulphur dioxide and ozone. Ozone resulting from the action of sunlight on exhaust fumes poses a major problem for asthmatics in cities like Los Angeles. Hair and deodorant sprays are usually avoided by those with asthma, since they readily irritate the airways and cause wheezing.

Air temperature and humidity   Frosty days can be a nightmare for asthmatics, though some people with this disease are worse in hot, humid weather. Quite why some are affected by cold and others by hot air remains to be explained. However, the reason why many asthmatics are worse after thunderstorms is now known and is due to release of the spores from a mould called *Didymella exitalis* described in detail in Chapter 3. Six hours after the start of the heavy rain up to 25 000 spores are found in every cubic metre of air, provoking asthma in those allergic to them.

Drugs   It has long been known that aspirin (the common name for acetyl salicyclic acid) can make asthma worse in about one in 20 of those suffering from the disease. Recently, other drugs which work in a similar way, and are used to control pain, have also been found to provoke attacks. A good example is Nurofen, which can be bought over the counter at the chemist's, a single tablet of which recently caused death in three asthmatic patients. All asthmatics must be careful about the use of these drugs, preferably avoiding them altogether. The painkiller which has caused fewest problems is paracetamol. Just as dangerous to asthmatics are the drugs known as beta-blockers, which are used for the treatment of high blood pressure and angina; examples are Inderal and atenolol. Their action is to block the effect of the hormone adrenalin which, as described on page 78, is so important in keeping the airways open.

What happens when you inhale an allergen?

Careful study of what happens in the lungs when an allergen is inhaled has helped us to understand the nature of asthma. Most asthmatics are well aware that an attack can start within minutes of contact with substances to which they are allergic. Measurements of lung function taken at such a time show substantial falls which, provided the contact is brief, improve within the hour. This is called an immediate asthmatic reaction and is due to the release of powerful chemicals from mast cells in the lining of the airways, described on page 14. These immediate attacks of asthma respond very well to treatment with bronchodilator aerosols. Such immediate reactions readily allow asthmatics to identify substances to which they are allergic and which they can avoid in the future. Allergens from domestic pets and horses are often identified as the cause of symptoms in these circumstances. We now know that the situation is often not as simple as this! If lung function is measured in the hours after the immediate response, a further, much more prolonged and severe response can be seen to occur in the airways of the lung. This is called a late asthmatic reaction. It usually begins 2–3 hours after the original brief exposure and gets progressively worse over the next 3–4 hours until improvement occurs. The pattern of changes in lung function recordings following a short period of contact with an allergen is shown in Fig. 16. The curious thing about the late asthmatic response is that it does not respond at all well to treatment with bronchodilator aerosols; in fact steroid treatment, either as an inhaler or in tablet form, is the only really effective treatment. Fortunately, late asthmatic reactions do not occur in all asthmatic patients or after every allergen exposure. Quite

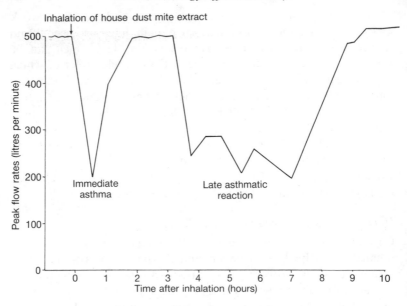

Fig. 16   Immediate and late asthmatic reactions.

why this should be remains a mystery, although we do know that asthmatics who show such reactions have worse disease than those who do not. In addition, it is now known that late reactions cause an increase in the responsiveness of the airways, making them more sensitive to future contact with allergens. Such a vicious circle means that continuing exposure to house dust mite, for example, can reduce a symptom-free asthmatic to one requiring urgent admission to hospital.

A 40-year-old Sussex arable farmer complained of a severe debilitating cough and shortness of breath beginning within minutes of shovelling grain and lasting for over a week. The fact that he still had symptoms on days when he was not in contact with grain misled his doctor who did not think grain allergy could be the cause of his breathing problems. Extensive tests in hospital showed just what was happening. After a week resting in hospital his symptoms disappeared and lung function tests were normal throughout the day and night. He was than asked to shovel for half an hour small quantities of the grain he had brought from his farm. This had a dramatic effect, as shown in Fig. 17. Within minutes of contact with the grain his peak flow fell precipitously and he experienced a severe immediate asthmatic reaction. After an hour or two he felt better but this did not last. His peak

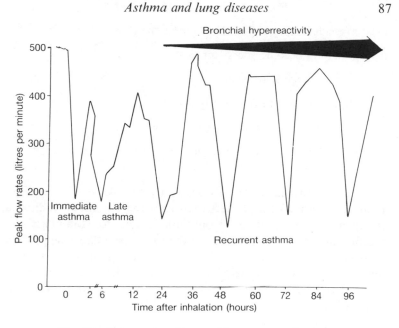

Fig. 17   Recurrent asthma and hyperresponsiveness.

flow fell again, but this time for much longer. By late evening he again felt better, but the next day in hospital, with no further contact with his grain, he developed a severe night-time attack of asthma. This recurred for the next four nights before steroid treatment was begun. What had happened was that his late asthmatic reaction had greatly increased the responsiveness of his airways giving him typical asthma with the diurnal changes described on page 78. In other words, a single, brief allergen exposure had led to severe and continuing asthma! It is not surprising that his doctor had found it so difficult to identify the cause of his asthma.

The complex changes that occur in the airways causing late asthmatic responses are described in Chapter 2. Basically, allergen breathed into the airways sets off a continuing inflammation, with the surface layers becoming damaged and eroded and the deeper regions becoming swollen and invaded by harmful cells.

How to recognize a bad attack of asthma

Fortunately, particularly with treatment, the vast majority of asthmatics rarely suffer bad attacks. More treatment is needed if:

(1)   breathlessness persists throughout the day;

(2)   coughing becomes persistent;

(3)   night-time waking occurs.

It is imperative to call the family doctor if:

(1)   it is difficult to complete sentences without gasping for breath;

(2)   continuing symptoms are no longer responsive to additional inhalations of bronchodilators.

The doctor may find a fast pulse, at a rate of over 100 beats per minute, and a peak flow of less than 150 litres a minute. Additional treatment, often including steroid tablets, is needed. Many asthmatics find that regular recordings of their own peak flows helps them to decide when their asthma is getting out of hand. Peak flow measurement can also help the patient and doctor to plan the treatment. For example, they can decide at what level of peak flow to call for help.

What happens when I go to hospital?

If the asthma is not responding to normal treatment, and has not been helped by the extra drugs given by the family doctor, admission to hospital will be necessary. Unfortunately, the need to go into hospital is becoming more and more frequent. An acute attack of asthma is now the commonest reason why children are admitted to hospital. This is true not only for the UK but for many countries in the developed world, particularly Australia and New Zealand, and seems to have become very common only in the past 10 years. This can be accounted for in part by:

(1)   more accurate diagnosis of asthma (many attacks were previously labelled as chest infections);

(2)   increased awareness of the need for prompt treatment.

However, these reasons cannot fully explain the dramatic increase in hospital admissions, and it seems likely that the disease has become both more frequent and more severe.

On arrival at the Accident and Emergency Department the doctor will ask questions about the asthma and the treatment that has been taken. If possible, all treatment should be taken to the hospital. This can be very important since, for example, it can be dangerous to give an injection of theophylline if the patient is already taking tablets of this drug on a regular basis. The action of this type of drug is fully described in Chapter 11. The danger from too much theophylline is that it can cause stimulation of the brain, leading to epileptic fits. During such fits breathing may cease for short periods, which is harmless in patients with no lung disease but can be fatal during a severe asthmatic attack. It is also very important to know whether injections or tablets of steroids are taken and whether a nebulizer has been used at home.

The doctor will then examine the patient's chest and check pulse and blood pressure. The peak flow will be measured and a sample of blood will be taken from an artery in the patient's wrist for measurement of the levels of oxygen and carbon dioxide. The level to which the oxygen in the blood has fallen and the carbon dioxide has increased are a very good guide to the severity of the asthma.

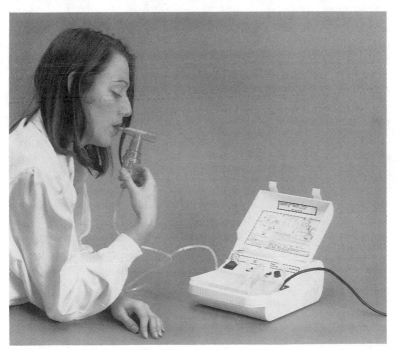

Fig. 18 Patient taking nebulizer.

After these measurements, treatment will begin, most commonly with nebulized bronchodilator, as shown in Fig. 18. Very occasionally acute severe asthma can be successfully treated in this way, and the peak flow will dramatically improve to perhaps as high as 300–400 litres per minute. In these circumstances a hospital stay may not be necessary, and the patient can go home, taking a course of steroid tablets. More frequently, this nebulizer treatment will only cause a small improvement in the peak flow and the patient will have to be admitted to the ward. On the way the patient will have a chest X-ray to make sure that there are no complications, the most common one being leakage of air into the space between the lung and the chest wall. This is called a 'pneumothorax' and is the result of high pressures generated in the lungs when trying to breath out through narrowed airways, which cause any weak part of the lung to burst.

Intensive treatment will be given on the ward. It is most important that an injection of a steroid (hydrocortisone) is given every 2–4 hours for the first day. At the same time, a large dose of steroid tablets in the form of prednisolone will be given. Nebulized bronchodilator treatment will continue every four hours. Provided progress is satisfactory, the frequency of nebulizer treatment will be reduced and hydrocortisone injections stopped. Most patients need to be in hospital for 4–5 days and require high-dose treatment with steroid tablets for at least two weeks.

Can asthma kill?

One of the reasons why asthma has not been treated as seriously as it should have been was the firm belief that asthma could not kill. We now know that this is totally wrong, and that over 2000 people die each year in the UK from asthma (Fig. 19). This is more than the number of deaths from illnesses which are considered so fatal, such as leukaemia. In some countries, deaths directly due to asthma have increased dramatically. Indeed, in New Zealand deaths have risen fivefold in the past decade. There is no good explanation for why deaths should be increasing. The peak of deaths which occurred in the 1960s was attributed to the introduction of bronchodilator aerosols which contained a substance called isoprenaline. This was very good at relaxing the muscles round the airways in the lung, but also stimulated the muscle of the heart. Many doctors thought that such stimulation of the heart in people whose lungs were working badly, and had reduced amounts of oxygen in the blood, was

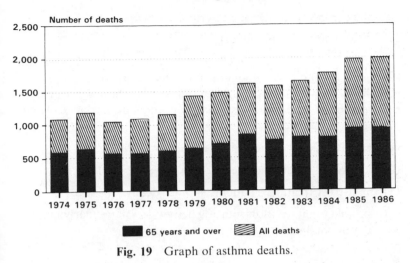

Fig. 19 Graph of asthma deaths.

sufficient seriously to disturb the pattern of heart beats, leading to heart failure and death. It is possible that this was true, but it is no longer a problem since modern bronchodilator aerosols, equally good at relaxing the muscles round the airways, have no effect on the heart. The most important factor that is found in the majority of cases of death from asthma is insufficient treatment with steroids. Every study of asthma deaths throughout the world has shown that inadequate treatment with corticosteroids is the most important factor leading to death. Often doctors, and sometimes patients, have underestimated the severity of their asthma and relied upon aerosol bronchodilators only when, at the very least, steroid aerosols and often steroid tablets were necessary. Patients are often reluctant to take steroids as they fear that there will be unacceptable or dangerous side-effects. These are discussed fully in Chapter 11, but it should be realized that the dangers from asthma which is inadequately treated and out of control greatly outweigh any possible side-effects of steroid treatment.

How to treat asthma
No one should suffer disabling symptoms from asthma provided that:

(1) the disease is diagnosed early;

(2)   the correct treatment is prescribed;

(3)   treatment is taken regularly

The approach to the treatment of asthma has changed radically in the 1980s. It is no longer thought sufficient simply to relieve wheeze with bronchodilator aerosols. The purpose of the treatment is to suppress the disease process so that symptoms do not arise. Treatment of asthma should:

(1)   enable the asthmatic to lead a full, active life;

(2)   keep lung function as near to normal as possible.

On the correct treatment patients should ideally:

(1)   be able to take part in sporting activity, so particularly important in children;

(2)   lose no time from school or work;

(3)   suffer from no night-time symptoms.

The achievement of such an ideal will, in almost every case, involve regular daily preventative treatment, usually with inhaled sodium cromoglycate, known as Intal, or steroids such as Becotide, Becloforte, or Pulmicort. Bronchodilator aerosols such as salbutamol, known as Ventolin or Salbulin, or terbutaline, known as Bricanyl, should be used if symptoms still occur in spite of preventative treatment. If it is necessary to use such bronchodilator aerosols more than 2–3 times a day, then insufficient preventative treatment is being taken. Bronchodilator aerosols are particularly useful in preventing asthma induced by exercise and it is wise for asthmatics, even though well-controlled on preventative treatment, to take two puffs of Ventolin or Bricanyl prior to strenuous, prolonged exercise. While the majority of asthmatics will find that this type of treatment controls their symptoms, a small proportion, probably fewer than one in 10, will require additional treatment with steroid tablets, such as prednisolone. The principle of treatment with these powerful tablets is to balance relief of symptoms and maintenance of good lung function against the side-effects which occur when prolonged oral treatment is given with high doses of steroids. In fact, side-effects are uncommon when the daily dose of prednisolone is less than 15 milligrams a day, and rare when the dose is less than 10 milligrams daily. At a dose of 5 milligrams per day side-effects are virtually unknown.

In the UK especially, doctors recommend the use of inhaled treatment whenever possible. This is because the doses needed for adequate treatment are very much lower when the drug is inhaled directly into the lung rather than swallowed. In addition, drug companies have devised steroids which either cease to be active as soon as they are absorbed from the lung into the blood or are rapidly broken down in the bloodstream. The major problem is that many asthmatics find it difficult to co-ordinate activation of a metered-dose inhaler with inspiration. The correct technique is fully described on page 173.

Many asthmatics who have this problem find it much easier to use a Spinhaler or Rotahaler where, once the capsule is broken, a slow steady inspiration is also required; this is explained in detail in Chapter 11. Nebulizers have proved popular for the delivery of bronchodilator treatment. Nebulizers supply an aerosol to a mouthpiece or face mask and allow the drug to be inhaled as a mist during several minutes of normal breathing. In fact there is little advantage in favour of nebulizers rather than ordinary Rotahalers or metered-dose aerosol inhalers, other than that the dose delivered in a nebulizer is up to 25 times greater. Nebulizers are expensive, and simpler devices are available to aid inhalation, for example 'spacers' on Bricanyl metered-dose inhalers or Nebuhalers, as illustrated in Chapter 11.

Asthma and pneumonia

Surprisingly, asthma is rarely triggered by the bacterial infections which can cause pneumonia, but asthma and pneumonia can occur together as a result of:

(1) inhalation of the spores of the mould *Aspergillus fumigatus* (allergic bronchopulmonary aspergillosis);

(2) some worm infestations, especially in the tropics;

(3) rare disease of unknown origin, such as polyarteritis nodosa.

Easily the commonest in the UK, although still rare, is allergic bronchopulmonary aspergillosis. As described in Chapter 3, this mould is unique in that it acts both as an allergen and as an infective agent actually growing in the airways of the lung. The characteristics of the disease that results are:

(1) attacks of wheezing and breathlessness;

(2)   severe cough with hard sputum;

(3)   damage to the large airways of the lung;

(4)   pneumonia.

Quite unlike bacterial pneumonia, this allergic pneumonia responds quickly and completely to treatment with tablets of prednisolone. Patients with this disease, like other patients with allergy, have positive results on skin-testing with extracts of *Aspergillus fumigatus*, and not only IgE antibodies but also IgG antibodies can be found in their blood. Fortunately, this manifestation of allergy to *Aspergillus fumigatus* is rare and it is much more common for spores of the mould to cause uncomplicated asthma.

Allergic pneumonia

Allergic reactions can affect the lung tissue itself rather than the airways, to produce a pneumonia-like illness, called allergic pneumonitis in the USA. In the UK, however, such diseases are described as allergic alveolitis to emphasize the involvement of the gas-exchanging parts of the lung, the alveoli, (see p. 69). Not surprisingly, the major symptom of these diseases is breathlessness. Farmers, particularly those in wet parts of the world, are most frequently affected. Indeed, in the wet, west-coast areas of the UK up to one in 10 farm workers suffers from 'farmer's lung'. The cause of this ailment is a micro-organism called *Micropolyspora faeni*, that multiplies in damp and mouldy hay. These microbes do not grow in hay until the water content is more than 25 per cent, and their growth causes heating of the hay—often up to 50 or 60 °C. When such mouldy hay is thrown down for cattle fodder and bedding, masses of bacteria are liberated into the air. Often this occurs in the winter-time when the barn is closed tight against the cold, and the farmer may inhale up to a million 'spores' each time he works in the barn. The tiny size of these particles allows them to go right through the airways into the alveoli where they stick and cause an allergic reaction. Surprisingly, the farmer does not notice a thing while inhaling the microbes, but hours later, in the evening, 'flu-like symptoms occur with aching joints, high fever, coughing, and breathlessness. Just as surprisingly, all symptoms have cleared by the next morning, but the process starts again if the farmer re-enters the barn to work with the hay. If the farmer does not get away from the barn, permanent damage will eventually occur causing severe shortness of breath and making it impossible for the farmer to continue his job. For this reason farmer's lung is recognized by the DHSS as a 'prescribed' industrial disease. This means that workers affected by the disease and unable to continue in employment can claim compensation and be paid a disablement benefit. Farmers thought to be affected are assessed by doctors serving on Medical Boarding Panels (in chest disease) who will award a pension if:

(1)   the symptoms are typical;

(2)   the chest X-ray and breathing tests show lung damage;

(3)   there is evidence of an allergic reaction to *Micropolyspora faeni*, or a closely related organism *Thermoactinomyces vulgaris*, by detection of antibodies in the blood.

If this disease is recognized early enough, permanent damage can be avoided provided the farmer is no longer exposed to mouldy hay, although this may mean giving up his livelihood. However, the frequency of this disease has been greatly reduced in recent years by modern farming methods, in particular the use of storage silos.

Unfortunately, these allergic pneumonias are not confined to farm workers and there are many other occupations in which workers can inhale large amounts of micro-organisms. For example in the cheese, lumber, and cork-making industries. In more tropical countries, workers handling damp cane fibre (bagasse) left after extraction of sugar from cane can develop exactly the same disease as farm workers. In each case a different micro-organism is responsible. In the past few years, with the widespread use of air-conditioning and humidifying systems for the maintenance of constant temperature and humidity, so essential for many modern manufacturing processes and computer systems, 'humidifier fever' has become a recognized disease. Once again, the features are identical to farmer's lung but can occur in any large building in which recycled water is used to humidify air. The tanks in which the water is stored readily become contaminated with a wide variety of minute living organisms, many of which have been shown to produce 'humidifier fever'. Once recognized, this problem can easily be cured by raising the temperature of the water or adding suitable disinfectants, though the latter may contribute to other symptoms which have been called the 'sick-building syndrome'. These examples underline how careful we must be in controlling our environment and ensuring that the workplace is safe. However, it is not just the work environment that can cause lung problems.

Everyone knows how easy it is to become allergic to pets, particularly to dogs, cats, and horses, but few people realize that the budgerigar sitting in its cage in the corner of the room can be a cause of progressive breathlessness and even death. Since almost one-quarter of the households in the UK keep budgerigars, it is fortunate that this disease, known as 'budgerigar fancier's lung', is very rare. The disease results from inhalation of dried particles from budgerigar droppings, so the disease mainly affects those who are responsible for cleaning out the cages. Although 'budgerigar fancier's lung' is extremely rare, 'pigeon fancier's lung' is not, due to the larger numbers and closer contact that the fancier has with his birds. The cause in this case is the inhalation not only of dried bird droppings but also the fine powdery 'bloom' which covers the feathers giving them their sleek and shiny appearance. One-quarter of pigeon fanciers have evidence of allergy to pigeons, but not all suffer the symptoms of allergic pneumonia. Those that do experience fever, chills, coughing, and shortness of breath several hours after handling their birds. Once again, cure is possible if the disease is recognized early and the hobby given up.

Sometimes avoiding the cause of these allergic pneumonias does not cure the disease—even in the early stages—and usually improvement will occur following treatment with steroid tablets.

7

ECZEMA AND RASHES

The surface area of the skin of an average-sized man is 1.6 square metres, making it the largest organ in the body. It is of great importance to survival since it forms a barrier between our bodies and the outside world. Without it we would be extremely susceptible to changes in temperature and weather, and also at great risk from infection. Indeed, it is impossible to survive if more than 60 per cent of the surface skin is lost through burns. The functions of the skin are numerous, but the most important are:

1. To protect against invasion by bacteria, viruses, parasites, insects, and worms.

2. To prevent both excessive loss of water from the body and absorption of water from the environment.

3. To protect against the ultraviolet rays of the sun.

4. To use sunlight to make vitamin D, preventing rickets.

5. To alert us to change in our environment through the sensations of touch, temperature, and pain.

The skin is divided into three layers. The topmost of these is called the epidermis, and consists of layers of cells which gradually become flatter and tougher and then die as they reach the surface. Each of us sheds up to 1 gram of these dead cells into our clothes and beds every day. This loss of cells is obvious when it becomes excessive in people suffering from dandruff (seborrhoeic dermatitis). The epidermis contains two important additional cells:

1. Dark, pigment-containing cells, called melanocytes, which give the distinctive skin colour of different racial groups and cause pale skin to go brown in the sun.

2. Cells which can engulf foreign materials, such as chemicals and allergens, processing them ready to trigger an immune response.

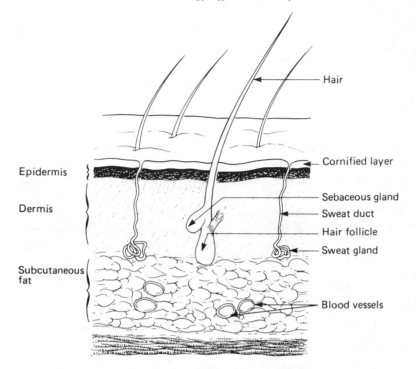

Epidermis

Dermis

Subcutaneous fat

Hair

Cornified layer

Sebaceous gland

Sweat duct

Hair follicle

Sweat gland

Blood vessels

Fig. 20 Structure of the skin: section showing skin layers.

These are called Langerhans cells after the scientist who first described them.

Below the epidermis is the dermis, the layer which consists of fibrous and elastic material which gives the skin its flexibility. Also in the dermis are blood vessels, nerves, sweat glands, and hair roots. Below the dermis is the subcutaneous layer, largely consisting of fat which insulates and contributes towards the characteristic male and female body shapes, females having more subcutaneous fat than males.

Few of us, if any, get through life without suffering some form of skin allergy, be it the development of a red, itchy, and scaly rash following contact with plants, chemicals, or jewellery, or the development of itchy bumps after eating strawberries or shellfish. One of the commonest problems in childhood, causing great distress to mother and infant alike, is a weepy, itchy rash in the skin crease of the elbows and knees. This is usually called atopic eczema, which is one type of dermatitis (meaning inflammation of the skin), the other being called contact dermatitis.

Atopic eczema

Almost one in 20 people suffers from this skin rash at some time in their lives. The adjective 'atopic' implies that there is an association between the skin rash and an allergic constitution. Atopic individuals are those who readily develop allergy to common substances around them. Other allergic diseases, such as asthma and hay fever, also occur in up to half of the children with this type of eczema. There is a hereditary predisposition: when both parents have eczema, there is a greater than 50 per cent chance that their offspring will suffer from the disease. There are three main types of atopic eczema:

● infantile eczema

● flexural or childhood eczema

● adult eczema

Infantile eczema

This usually starts between the ages of two and six months. The weepy, scaly rash starts on the face and the irritation keeps the baby awake. The child may rub his face or sometimes his scalp on a pillow or cot side in an attempt to relieve the itching. The nappy area is not involved. The rash clears in less than half of the infants by 18 months of age. As the child gets older the pattern of the disease changes.

Flexural or childhood eczema

From 18 months onwards the rash typically affects the skin folds, particularly the inside of the elbows and the back of the knees. Additionally, the skin on the side of the neck, wrists, and ankles is affected. In toddlers the eczema often persists on the face.

Adult eczema

Atopic eczema is less common in adults. The rash principally affects the skin folds where thickening may occur. Emotional problems can sometimes prolong the course of the disease, but it rarely persists above the age of 30. Localized patches of eczema can occur on the lips and sometimes on the nipples, especially in young women. Recurrence of eczema may occur on stressed parts of the body, for example on the hands of nurses and hairdressers. Eczema affecting the hands or feet often appears as small blisters under the skin. In older people, patches of eczema are often seen associated with varicose veins around the ankles.

Surprisingly little is known about the cause of this common disease. Most sufferers show positive results on skin-prick testing with common inhalant and food allergens, as is the case in people who have extrinsic asthma. Likewise, the majority show raised levels of immunoglobulin E in their blood. One modern theory suggests that eczema is a result of contact between damaged skin and allergens from the house dust mite or domestic pets. For poorly understood reasons this causes the typical change of eczema rather than the characteristic raised weals of urticaria. Diet is important in causing eczema. Exclusive and prolonged breast-feeding can delay, and in some cases prevent, the disease. The most important food allergens are milk and eggs. Hot and humid weather, drying of the skin, and woollen clothes can all cause eczema to flare up.

Contact dermatitis

The skin rash in contact dermatitis looks exactly like atopic eczema, although the red, scaly and itchy patches occur anywhere on the body where there has been contact with the causative material. Indeed, the site of the rash usually indicates the likely cause. Contact dermatitis is caused by:

● irritation
● allergy

Irritant dermatitis

When the skin loses the oils and water that keep it moisturized, dryness and cracking occur. Up to one in five housewives are affected at some stage. The severity of this depends on the frequency with which the hands are immersed in water when washing-up or cleaning. The problem often starts beneath a ring where neat detergent or soap tends to lodge, causing dermatitis which often spreads to affect the thin skin on the sides of the fingers, the webs between the fingers, and the back of the hand. In severe cases even the palms can be affected.

Irritant dermatitis is particularly common at workplaces where solvents used to remove grease from metal surfaces equally easily remove oils and water from the hands. These problems can be avoided in the home and at work by wearing suitable protective gloves, though these are known to produce allergic contact dermatitis in a few unfortunate people.

Allergic contact dermatitis

Simple chemicals are the commonest cause of this disease. Such materials combine with protein in the skin and this combination starts an allergic reaction in some people. Atopic people are not at any greater risk of developing contact dermatitis than their non-atopic colleagues, but the incidence of this disease is particularly high in certain industries. For example, one in 10 bricklayers and carpenters suffers from this problem caused by cement or wood resins. For reasons that are not understood, the skin of some people is more easily sensitized than others. Common sites and sensitizers are:

1. *Scalp.* Most commonly affected by hair dyes, or by shampoos, conditioners, or mousses. The scalp is resistant to sensitization and the dermatitis affects the skin along the hair line.

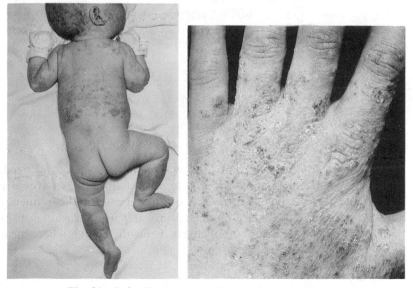

Fig. 21 Infantile eczema and contact dermatitis.

2. *Face.* Cosmetics and aftershave lotions. Spectacles made from nickel or even plastics can cause dermatitis where they are in contact with the nose, cheeks, and eyelids.

3. *Eyelids.* Eye shadow, mascara, and hair spray. Nail varnish and plant allergens can be deposited when the eyes are rubbed.

4. *Lip area.* Lipsticks, toothpaste, gargles and mouthwashes, and nickel from lipstick cases which rub against the lips when the container is nearly empty.

5. *Ears.* Perfume, stereo earphones, telephone receivers, and ear-rings. Ear-rings are now the commonest cause of nickel allergy, which used to be due mainly to suspender fastenings. Nickel sensitivity rises sharply in girls between 12 and 16 years of age when the wearing of jewellery becomes commonplace.

6. *Neck.* Perfume, metal from jewellery, and garments. Nickel from the clasp of a necklace of from a zip fastener may result in a small spot of dermatitis at the nape of the neck.

7. *Armpits.* Deodorants, depilatories, and garments.

8. *Hands.* The most important site, accounting for two-thirds of all contact dermatitis. Many possible causes including hand lotions, rubber gloves, jewellery, plants, and occupational materials.

9. *Body.* Nickel from jeans studs and zip fasteners. Leather, plastics, and rubber in garments.

10. *Anus.* Medications for haemorrhoids (piles). In very rare cases dyes and perfumes in toilet papers can be involved.

11. *Genitals.* In women problems can be caused by contraceptive creams or jellies, rubber diaphragms, condoms, feminine deodorants, douche additives, vaginal medications, menstrual pads, or tampons. In men dermatitis in the genital area can be caused by condoms, contraceptive creams, or vaginal medications used by a sexual partner. Isolated patches on the upper thighs can even be caused by objects kept in trouser pockets.

12. *Feet.* Shoes, slippers, and athlete's foot remedies. Rubber boots or wellingtons cause problems where their upper edge is in

contact with the lower leg. Socks may affect the lower leg due to rubber in their elastic welts.

Urticaria—nettle rash

We are all familiar with the itchy white blisters and red patches that result from nettle stings. This closely resembles urticaria which, like the nettle sting, is mostly due to histamine. Typically, urticarial rashes come and go over the space of a few hours, on any part of the body. This condition is extremely common, affecting up to one in five people at some stage. Whereas inhaled allergens, such as the house dust mite, cause rhinitis and asthma, food allergens often cause urticaria, though some allergens can cause both. Animal allergens, when inhaled, cause asthma, but saliva from dogs and cats licked on to the skin can cause urticaria. People who are allergic to grass pollen can develop urticaria on the legs while walking through long grass or sitting down for a picnic.

Urticarial rashes can be short-lived, lasting a few hours or days, or persist for several weeks, months, or even years. Short-lasting urticaria is usually caused by:

(1) seafoods, particularly shellfish, such as crab, prawns, and lobster, and bony fish, such as cod and mackerel;

(2) berry fruits, particularly strawberries;

(3) nuts, particularly peanuts, cashew, and hazel-nuts;

(4) eggs;

(5) chocolate;

(6) medicines, particularly antibiotics, such as penicillin, though almost any medicine can be a cause;

(7) plants—nettles and strawberry leaves;

(8) pollens;

(9) animals;

(10) house dust mites;

(11) insects—bees, wasps, and hornets; or

(12)   worms—only those that infest the body.

While many of the common causes of urticaria involve allergy, some can cause the rash in people who are not allergic. Some foods such as old cheese and tuna fish contain very large amounts of histamine which, when eaten, cause the rash directly. Other foods, particularly strawberries, although not containing histamine, do contain substances which cause release of histamine from mast cells by a mechanism which does not involve allergy. Obviously, it is very difficult to distinguish such reactions from allergy. This really does not matter since the treatment is exactly the same—avoiding the food.

Some types of urticaria result from entirely different causes. This is especially the case in those with very sensitive skins (one in 20 people), where a slight scratch which does not break the skin causes redness and blistering. The medical name for this is dermatographism, and the urticarial rash is the same shape as the original scratch.

Urticaria can develop in some people:

(1)   when they get particularly hot or cold;

(2)   on sweating or after a bath;

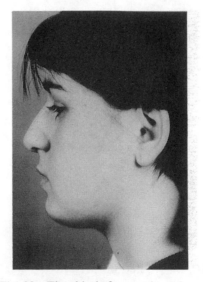

Fig. 22 The skin before angio-oedema.

(3)   when they are nervous or embarrassed; or

(4)   on exposure to the ultraviolet rays of the sun.

Urticaria that lasts for more than six weeks has a more complex cause, usually unrelated to allergy. Often the rash is associated with other illnesses, particularly infections by viruses. Very rarely an underlying cancer is the cause. Strangely, such long-lasting urticaria is often kept going by aspirin and related chemicals such as food colourings. Avoidance of these allows the rash to settle.

Angio-oedema

This complex name describes urticaria occurring in the deeper part of the skin and underlying fatty tissue which becomes very swollen. All areas of the body can be affected. When angio-oedema affects the face it totally changes appearance, as is shown in Figs. 22 and 23. If it affects the tongue, throat, or vocal cords the swelling can prevent air getting into the lungs and can cause death unless promptly treated. Fortunately, angio-oedema is very rare. Like urticaria it can be caused by allergy, but is more often provoked by aspirin and

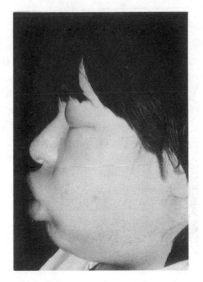

Fig. 23   The skin after angio-oedema.

related compounds. Very occasionally it runs in families, and the frequent attacks require prolonged treatment with hormones or drugs which affect blood clotting.

The itch

Intense itching can result from allergy to mites which, unlike the house dust mite, can burrow into the skin. The proper name for these mites is *Sarcoptes scabiei*, and the disease is scabies. Typically the mites tunnel into the skin between the fingers, and around the wrists and ankles. Tiny burrows can be seen with a very small blister at the end from which the mite can be picked out on the end of a needle. Scabies is passed from one person to another by close contact, usually in the warmth of the bed. The itching, caused by allergy to the mite, its eggs, and droppings, is so severe that the skin is often scratched and gouged. Treatment requires laundering of all clothes and bed linen and painting of the entire body, except the face, with a disinfectant lotion called lindane. All members of the family must be treated in this way. Fortunately, improved standards of personal hygiene and better living conditions have made this disease rare in the developed world.

Sunlight and the skin

People often say that they are allergic to sunlight when rashes, itching, and blisters spoil a day by the seaside. In fact, moderate amounts of sunlight can often help to clear many skin conditions, in particular eczema, acne, and psoriasis. In addition, sunlight acts on chemicals in the skin to produce the essential vitamin D without which bones do not grow properly. Unfortunately, too much sunlight, especially in fair-skinned people, can cause burning and, over longer periods, premature ageing of the skin with permanent wrinkling, pigmented patches, and cancer.

Some people are indeed more sensitive than others to sunlight and can develop red, itchy patches on exposed skin, particularly in spring and early summer. The fair-skinned are especially prone to this photosensitization but, fortunately, the rash improves as the summer progresses and the skin hardens. Sun-blocking creams with high protection factors can help, as can steroid preparations.

Unfortunately some perfumes, cosmetics, and even sun lotions contain chemicals, usually preservatives and antiseptics, that make the skin more sensitive to sunlight. Occasionally even prescribed drugs, for example the sulphonamide and tetracycline groups of antibiotics, some diuretics, and tranquillizers, can have the same effect. Such 'photosensitizers' damage the skin in one of two ways:

1. Phototoxicity—by concentrating ultraviolet light from the sun in the skin, causing more rapid burning.

2. Photoallergy—by reacting with substances in the skin which trigger an allergic reaction when exposed to sunlight, even when filtered through glass, resulting in urticaria.

The leaves of some plants, for example giant hogweed and wild parsley, contain chemicals called psoralens which concentrate light and, if rubbed into the skin, cause phototoxic burns.

Acne

Most of us suffer from blackheads and spots on the face and back, particularly in adolescence. Usually they are transient and, apart from embarrassment, cause little problem. In some people, however, acne becomes extensive and persistent. No one is clear about the cause of this common problem but food allergy has been suggested. It is more likely that the spots result from hormonal imbalance and the activity of a bacterium called *Propionibacterium acnes*, which lives in glands around hair roots. These glands secrete a greasy substance called sebum which oils the hair. Increased bacterial activity causes the duct of the gland to become blocked and pigmentation from the hair causes the typical blackhead in the skin. Infection then progresses to give a red spot. Treatment is based on the use of substances that cause mild peeling of the skin, such as benzoyl peroxide—a bleach—contained in preparations such as Acnegel, or sulphur in Dome Acne. Exposure to sun and wind has the same effect. If such treatment is insufficient, regular antibiotics, such as tetracycline, may be helpful. Daily treatment with an oral contraceptive, containing a substance which blocks the action of male hormones which promote sebum secretion, can be very helpful in young women. One such preparation containing cyproterone acetate and ethinyl oestradiol is called Diane. In very severe cases, treatment with a substance called isotretinoin, which shrinks the sebum glands, can have dramatic effects. Unfortunately the drug has substantial side-effects, including dry lips, sore eyes, nose bleeds, hair loss, and joint pains, which restrict its use.

There is no evidence that allergy causes acne. However, some people find that diets low in fats and avoidance of chocolate help, though why this should be is not known.

Rashes caused by drugs

Side-effects due to drugs are common and most often affect the skin or digestive system. The frequency with which they occur increases with age, affecting less than one in 30 adolescents, but one in five old people over 80. As mentioned in Chapter 3, drugs produce unwanted side-effects in several different ways, the most important being intolerance and allergy. It is difficult to appreciate the difference between intolerance and allergy since both mechanisms can produce similar symptoms. In allergy (e.g. penicillin causing a skin rash) the person does not react to the substance on first meeting it, and very specific immunological mechanisms involving antibodies are involved. In intolerance (e.g. aspirin triggering urticaria) no previous contact is necessary, and specific immunological mechanisms are not involved.

Allergy is the commonest mechanism, causing rashes due to drugs (which typically are small chemicals which must react with protein in the body before they can trigger an allergic reaction).

Drug allergy can cause a whole array of different skin reactions, from simple itching to widespread eczema. Bruising, spots, itchy red patches, and blisters can all result from allergic reactions to the simplest drugs, including those bought over the counter from the chemist. Most such rashes require no treatment and usually clear within three to four days after the offending drug is stopped. Drugs applied to the skin are an important cause of allergic dermatitis, which can be particularly confusing when antihistamine creams are applied to itchy spots, since allergy can develop to the antihistamine itself, though more often to stabilizers, such as ethylenediamine, present in the cream. This also applies to some steroid creams so useful in the treatment of skin problems.

Well-known drugs causing rashes include:

(1) antibiotics, particularly penicillins, sulphonamides, and tetra-
cyclines;

(2) painkillers and migraine remedies, particularly aspirin, codeine, and ergotamine;

(3) sedatives, particularly chlorpromazine;

(4) antihistamines and local anaesthetics, particularly when applied to the skin.

It is very important to realize that although some drugs, such as those in the list, produce rashes fairly frequently, there are an enormous number of drugs which can, on rare occasions, cause skin trouble. On this basis it is wise to consider the likelihood of a drug allergy as the cause for any rash; even oral contraceptives can be a rare cause of photosensitivity.

8

FOOD ALLERGY

There is hardly a more controversial issue in medicine than food allergy. Allergy to food has been blamed for almost every conceivable symptom from diarrhoea to mental illness. Increasingly, people are both conscious of and concerned about what they consume—after all, 'we are what we eat'. The current fashion of selling 'organic' or 'preservative and colouring free' foods reflects public concern, but what are the facts? It is now clear that only a minority of unpleasant reactions to food are truly due to allergy. In most cases this may not matter, since whatever the mechanism causing the symptoms, the cure is the same—avoidance.

Food allergy illustrates beautifully the complex way in which we respond to the fundamental need to eat. Vegetarianism is just one example of how our minds can influence what we eat. However, food avoidance or aversion taken to the extreme can prove fatal, as in anorexia nervosa. Everyone knows that too much coffee can prevent sleep. This is not an allergy but a direct effect of the stimulant caffeine. Migraine is often brought on by eating foods such as bananas and cheese, which contain large amounts of a natural chemical called tyramine which can affect the blood vessels in the brain. Again, this is not an allergy but rather an intolerance present in a minority of people. On the other hand, there is no doubt that other foods, such as milk, eggs, and nuts in particular, can, in some people, cause the most severe forms of allergic disease.

Unpleasant reactions to foods are best considered as either being controlled by the mind (food aversion) or due to a physical cause (food intolerance). Although there is no doubt that such reactions are very common, no one knows just how frequent they are.

Food aversion

Our need for food varies throughout life. Adolescence is a time of growth, requiring a high calorie intake. However, girls much more frequently than boys consider that they are overweight, and dieting, often ineffective, is widespread in young women. Food likes and dislikes are common, and refusal of food in infancy is usually part of the child's drive for independence. However, in later life overeating or rejection of certain foods (food faddism) is usually the result of psychological factors which are not clearly understood, though they may be the result of dietary patterns established in childhood. Clearly, parents have a great influence on their children's ideas about food. This is harmless when confined to high-fibre and organic foods, the so called 'muesli belt' children, but can be pernicious if it leads to total exclusion of certain foods to which the child is then often said to be allergic. This is best described as food intolerance by proxy.

The most severe form of food aversion is anorexia nervosa, characterized by a determined refusal to eat, which usually begins in adolescence. Malnutrition can slow the onset of sexual maturity. Patients with anorexia stay thin either by refusing to eat food, particularly foods high in carbohydrates, by inducing vomiting shortly after eating, or by taking large doses of laxatives. Anorexics try to hide their condition from family and friends by maintaining that particular foods are harmful or that they are allergic to them. Often such people have a deep-rooted interest in food, but usually prepare and eat it alone. Anorexics suffer from lethargy with occasional bursts of frantic activity. They often complain of many bodily symptoms, such as asthma, migraine, faints, fits, joint aches, stomach pains, and even kidney trouble. It seems likely that the condition known as the 'total allergy syndrome' is a form of anorexia nervosa occurring in people who are very manipulative.

Some people suffer from a related condition called bulimia in which there is an overwhelming desire to eat, particularly carbohydrate foods. The fatness that would result from this is prevented by induced vomiting, purging, and periods of enforced starvation. Bulimia is a private affair kept secret from family and friends, even though on binge days up to three times the normal calorie intake is consumed.

Mild depression itself is often characterized by overeating, although when the condition becomes more severe the appetite is usually lost.

Food intolerance

Food intolerance is a reaction to a specific food or food ingredient which is not affected by psychological factors. The reaction occurs even when the person concerned is unable to identify the food, for example when milk protein is disguised by its presence in white bread. Food allergy is one type of food intolerance which occurs when the immune system is involved in the reaction. Food intolerance includes:

(1) normal reactions to large amounts of specific foods, for example agitation following excessive amounts of caffeine in strong coffee (or tea);

(2) irritant effects, such as diarrhoea after eating highly spiced curry;

(3) digestive problems due to lack of chemicals, called enzymes, necessary for the breakdown of certain foods ready for absorption into the body, for example milk causing diarrhoea in people who have low levels of the enzyme lactase; and

(4) allergy.

The purpose of the digestive system in man (known as the gastrointestinal tract) is to break down food into its components which can then be readily absorbed into the body. The stomach acts as a reservoir for food, mixing it with acid and breaking down fats as a preliminary stage in its digestion before it enters the small intestine. In the small intestine the area available for food absorption is greatly increased in three ways, as shown in Fig. 24. From the small intestine food remnants move into the large intestine, comprising the colon and rectum, whose function is to absorb much of the remaining salt and water before the remnant is passed from the body.

Normal reactions to large amounts of specific foods

A cup of tea contains up to 80 milligrams of caffeine and a cup of coffee up to 150 milligrams, with lesser amounts in cola drinks. Caffeine is the most popular and widely used stimulant drug in the world, and produces side-effects at doses of only 200 milligrams, two moderately strong cups of coffee. Caffeine is addictive and particularly affects the nerves and the heart—though it has a useful action,

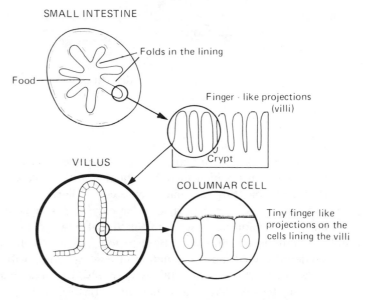

Fig. 24 Food absorption in the small intestine.

stimulating urine production and widening the airways of the lung. Too much caffeine produces an anxiety state with shaking of the hands, sweating, racing of the heart, and rapid breathing and insomnia. In addition, caffeine can precipitate migraine.

Chemicals which affect blood vessels are found in many foods. High concentrations of one such chemical, tyramine, are present in some cheeses and pickled fishes. A similar chemical, called phenyl-ethylamine, is found in chocolate, and citrus fruits contain octopa-mine. These and similar compounds also occur in other foods, such as bananas, avocado, yeast extract, and wine. Typically, these chemicals precipitate migraine in susceptible people and can also cause flushing of the face and urticaria. Some foods contain or are capable of releasing histamine in the body, producing symptoms indistinguishable from allergic reactions—so-called pseudo-allergic reactions. Histamine is a normal constituent of some cheeses, sauerkraut, and sausages such as pepperoni and salami. Many bacteria produce histamine and many of them grow well in stored fish, especially those of the Scombroid family, such as mackerel. Scombroid fish poisoning causes urticaria, angio-oedema, facial flushing, and intense headaches. Poisoning rather than allergy is the cause of such symptoms, since several people are likely to have developed symptoms simultaneously.

Irritant effects

Stomach-ache, wind, constipation, and diarrhoea are common complaints. Up to third of us frequently suffer from these troublesome symptoms, but we rarely complain of them to our doctors. Nevertheless, the occurrence of stomach-ache and constipation or diarrhoea in the absence of any serious underlying cause, such as ulcers or cancer, is called the irritable bowel syndrome. Because of the variety of symptoms, which include difficulty with swallowing, heartburn, nausea, bloating, stomach-ache, colic, diarrhoea, and constipation, this condition has also been called spastic colon, nervous diarrhoea, colon neurosis, and functional bowel disorder. Certainly, symptoms are influenced by the food we eat, which can alter the tenderness of the gut, production of gas, and composition of the faeces, but whether or not allergy is involved is difficult to prove. Most doctors, unlike many patients, consider that food allergy is only rarely a cause of the irritable bowel syndrome. Whether allergy is involved or not, those suffering from this disorder should omit foods that upset them, while making sure that they maintain a balanced diet. Many sufferers find that a high-fibre diet helps to relieve symptoms. Coping better with the stresses of life is

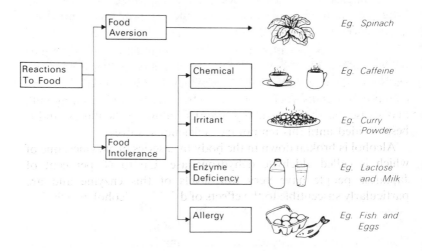

Fig. 25 Reactions to food.

also beneficial. It is self-evident that any irritant food, such as hot, spicy dishes, will make matters worse.

Inflammatory bowel disease

There are two diseases that cause severe inflammation in the bowel, called ulcerative colitis and Crohn's disease. These serious diseases can cause obstruction to the passage of food through the gut, with severe colicky pain in the stomach or copious bloody diarrhoea. In spite of much research, the cause of these diseases is unknown and eventually large lengths of bowel have to be removed by surgery. Some doctors think that allergy to milk plays an important part in these diseases, though this has never been proved. Nevertheless, elemental diets have been shown to be helpful, particularly in Crohn's disease. These contain all the necessary nutrients but in broken-down form. An example is 'Vivonex', which contains sugar, broken-down protein, and safflower oil, together with vitamins and minerals to which water is added before eating (see Chapter 11). Whether these diets work by resting the bowel, changing the bacterial content, or reducing allergy in the bowel is as yet unknown.

Enzyme deficiencies

Intolerance of certain foods may, very occasionally, be due to a lack of the enzyme needed for their digestion. The commonest is lack of the enzyme called lactase. Children born without this enzyme develop severe watery diarrhoea when fed with milk. Unrecognized, this condition leads to collapse and death. After childhood, milk is not so important in the diet and levels of lactase normally decrease. In some races, for example the Chinese, where milk plays little part in the diet, levels of lactase are extremely low and drinking milk leads to diarrhoea. Similarly, lactase levels are reduced by many common bowel disorders, such as gastroenteritis, and drinking milk may cause diarrhoea for many months following the illness, and is best avoided until this temporary deficiency resolves.

Alcohol is broken down in the body by a series of enzymes, one of which is called aldehyde dehydrogenase. Up to 40 per cent of Japanese people have very low levels of this enzyme and are particularly susceptible to the effects of drinking alcohol, which they learn to avoid.

Food allergy

It is surprising that food allergy is relatively rare and certainly far less common than allergy to inhaled materials, such as grass pollen, since only tiny quantities of allergen are inhaled into our nose and lungs whereas large amounts are eaten. Although cooking and digestion can reduce the potency of food allergens, the human digestive system needs to protect the rest of the body from large amounts of potential allergens in our diet while absorbing necessary nutrients. The ways in which allergy is prevented include:

1. Prevention of absorption—the lining of the gut has a slimy coating which contains special antibodies called immunoglobulin A (IgA) which react with foreign proteins in the diet making them non-allergenic.

2. Tolerance—the body can adapt to continuing exposure to many allergens. For reasons that are not clear food allergens in particular can bypass the immune response.

Food allergy is far more important in children than in adults. Many foods have been reported as causing allergy, and are described in Chapter 10. However, milk, eggs, fish, and nuts are the most important. When allergy to these foods is severe the first mouthful causes dramatic swelling of the lips, tongue, and throat, which can be life-threatening. Occasionally sufficient food allergen is absorbed into the bloodstream to cause a widespread allergic response called anaphylaxis, in which the development of severe asthma is combined with a rapid fall in blood pressure. Unless treatment is immediately available death ensues. Less severe allergy causes typical symptoms in the digestive system, especially colic and diarrhoea. Very occasionally food allergy can cause asthma and nasal symptoms.

Coeliac disease

One good thing to come out of the Second World War was the realization that some sickly children with persistent diarrhoea actually improved as their diet became more restricted, since they were unable to get bread. Eventually it became clear that a constituent of wheat, called gluten, severely damaged the lining of their digestive system. The villi, illustrated on page 113, were completely destroyed, so that absorption of food was dramatically reduced. Fat, in particular, passed straight through the body, and with it many vitamins essential for blood formation, resulting in thin, weak, anaemic children. Removal of gluten from the food (a gluten-free diet) totally transformed the lives of these children. Although the exact way in which gluten damages the lining of the gut is not fully understood, it is now known that an immunological response is involved, with antibodies reacting with a component of gluten, α-gliadin, in the lining of the gut.

Food additives

A food additive is any substance not commonly used as a food which is added to food at any stage to affect its keeping quality, texture, consistency, taste, or smell. Additives are described in detail on page 155–7. Typically, additives cause urticaria, asthma, and rhinitis, and only occasionally upset the digestive system. The most important colouring causing symptoms is the yellow dye tartrazine (E102). Apart from being added to so many convenience foods, it is usually present in smoked cod and haddock, lime and lemon squash, salad cream and marzipan. Surprisingly, tartrazine is a common constituent of mint sauce and jelly, brown sauce, and tinned processed peas. Tartrazine is a common colouring in many pills and capsules. Another yellow colouring, sunset yellow (E110), present in orange squash, fish fingers, and lemon curd, is also found in Lucozade and many medicines.

The chemical structures of tartrazine and sunset yellow are similar to each other and closely related to benzoates, other azo-dyes, and salicylates, including aspirin. This means that if a person has problems from one of these substances, reaction to the others will also occur. Salicylates are widespread in food, some naturally occurring and some artificially added. They occur naturally in a number of vegetables and fruit and are stable, appearing unchanged in jams and wines. A range of synthetic salicylates are used to flavour sweets, ice creams, soft drinks, and cake mixes. Benzoates (for example sodium benzoate (E211)) are preservatives, preventing the growth of bacteria and fungi, and are added to many foods, for example soft drinks, squashes, sauces, beer, and margarine.

Hyperactivity

Some children who are over-active have poor concentration, impulsive behaviour, and are difficult to control. Often they require only a few hours of sleep. This hyperactivity results in underachievement at school and disruptive behaviour. Many people, particularly in the USA, think that this is due to food additives in the diet. Doctors in the UK recognize that hyperactivity in relation to food does occur, but usually find that it is associated with urticaria, diarrhoea, migraine, and eczema. Nevertheless, Dr Feingold in the USA has developed a dietary treatment for such children. This treatment excludes food additives and has claimed considerable success.

Mental disturbance

There is no doubt that diet can affect mood. Food constituents can affect the activity of the brain, for example high-carbohydrate, low-protein meals elevate the levels of chemicals in the brain and influence activity. More severe mental illness, such as schizophrenia, is more common in patients with coeliac disease, leading to the idea that lack of dietary components, due to poor absorption, could be related to the onset of mental illness.

Migraine

The vast majority of headaches are due to tension in the muscles of the scalp and are often associated with the feeling of a tight band round the head, pressure behind the eyes, and a throbbing or bursting sensation in the skull. One in 10 of the population suffers fom recurrent headaches, often one-sided, and associated with changes in vision and upsets of the digestive system. Disturbance in vision usually occurs 15–60 minutes before the headache, which is frequently associated with nausea and vomiting. Sufferers are often irritable and prefer to lie in a darkened room. For unknown reasons, migraine often occurs at weekends, is common around puberty, the menopause (change of life), and pre-menstrually, and in those taking the contraceptive pill. Many find that certain foods can bring on an attack, especially chocolate and cheese, which contain large amounts of chemicals that affect blood vessels. Recently, doctors have shown that intolerance to food is important in the development of migraine in many children, and that exclusion diets can cure their symptoms. Allergy to foods may be the cause of migraine in these children since many also had eczema and complained of stomach-aches. They also reacted to many common allergens, such as grass

pollen and house dust mite, as did other members of their families.
Most attacks of migraine respond to treatment with simple
analgesics, such as paracetamol, though occasionally additional
therapy, such as metoclopramide, may be necessary for nausea.
These two drugs are often prescribed together as in Paramax.
Alternatively, it is possible to abort the attack if tablets of ergotamine are taken before the headache is fully developed. If migraines
occur frequently, treatment with antidepressant drugs, such as
mianserin, or drugs affecting blood vessels, such as propranolol,
may be useful.

Joint pains

It has been known for centuries that foods rich in protein, together
with alcohol, can cause gout, a type of arthritis. Aches in the joints
sometimes accompany urticaria and can be triggered by foods,
especially milk, but the role of food allergy in severe long-lasting
arthritis, such as rheumatoid arthritis, is not clear.

Part III

HOW CAN ALLERGY BE INVESTIGATED, DIAGNOSED, AND TREATED?

9

TESTS

The importance of recognizing allergies is that identification can lead to improvement and sometimes cure. The diagnosis of allergies depends on a combination of three factors:

(1) what the doctor is told about the symptoms;

(2) what the doctor finds when he examines the patient; and

(3) the results of special tests.

Tests on their own are not enough to diagnose allergy, but when combined with what the patient and doctor know about the symptoms allergy can be diagnosed.

Tests for allergy diagnosis

● skin tests
● breathing tests
● blood tests
● nasal tests
● food challenge
● X-rays

Skin tests

Many are mystified when it is suggested that they have a skin test to help diagnose allergic rhinitis or asthma. The reason is simple. Although the skin is not directly involved in such diseases as asthma and hay fever, it contains mast cells which, just like those in the nose or lung, are primed to react to allergens. Skin allergies, such as dermatitis and urticaria, are diagnosed using patch tests. Positive patch tests are the result of infiltration into the skin of lymphocyte cells in response to the application of the causative substance, e.g.

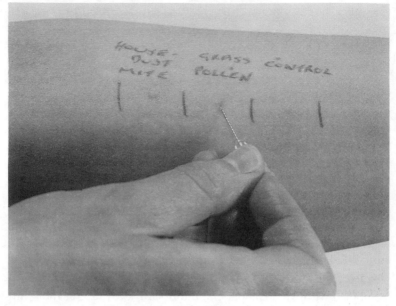

Fig. 26 Skin-prick testing.

nickel, whereas positive skin-prick tests result from the release of histamine and other chemicals from mast cells.

Skin-prick testing   This involves making a tiny prick with a needle through a drop of allergen placed on the skin. Testing is usually carried out on the inner surface of the forearm, although in infants it may be performed on the back. A number of allergens will be tested at the same time. The skin is marked with a pen to identify the site of each test and a separate needle is used at each site, so that there is no contamination by other allergens being tested which may cause a test to be falsely positive. Once a tiny amount of allergen (only about 100 000 molecules) has entered the skin at the prick test site it will trigger a reaction in any sensitive mast cells which it meets. Chemicals from these cells cause fluid to leak out of the tiny blood vessels in the skin forming a raised blister-like 'weal'. From the patient's point of view, a positive reaction will first be noticed as intense itching which begins within 1–2 minutes. This is followed by the weal, which slowly expands, reaching its maximum size in 10–15 minutes. The weal is surrounded by a wider area of redness and itching. Although the itching causes a little discomfort, the test itself is painless and is quite safe, since such a small amount of allergen is introduced into the skin that there should be no risk of even the most sensitive patient having a nasty reaction. The safety and simplicity of skin-prick testing have made it the method of choice in Britain and in many other countries.

Intradermal testing Intradermal testing is widely used in the USA. In this method a tiny amount of allergen is injected into the skin using a fine needle and syringe. Although very dilute allergen extracts are used, side-effects can occur in very sensitive patients. In addition it is painful, requires skill in its execution, and is liable to produce false positive results. For these reasons it is used only occasionally in Britain. The type of reaction on the skin which is produced by intradermal testing is very similar to that produced by the skin-prick test.

The initial reaction to both types of test disappears within 1–2 hours. However, an ill-defined area of swelling sometimes arises at the site of the test after 5–6 hours. This is a late skin reaction. Occasionally, a further reaction will occur on the day following a test. These delayed reactions are not dangerous; they are just the end-result of a process of inflammation started by chemicals released from the mast cells.

What do the results of skin-testing mean? Many people think that all the allergens to which they have a positive skin test are direct causes of their disease. This is not the case. Positive reaction means that the mast cells in the skin of the patient are sensitized to that particular allergen, but whether someone suffers from hay fever, asthma, or skin allergies depends on other factors. Indeed, some people have positive skin tests to grass pollen or house dust mite without ever noticing any symptoms. Skin testing alone cannot identify the cause of a patient's symptoms. The doctor needs to know all the details of the illness which, combined with the results of skin-testing, can identify the cause of the problem.

When Mrs A brought her seven-year-old daughter, Caroline, to the Allergy Clinic she was expecting to hear bad news. Caroline had been suffering from a blocked nose and bouts of sneezing for almost two years. Her symptoms were no worse during the hay-fever season but persisted throughout the year. A neighbour had suggested that Caroline might be allergic to the family cat and her mother was expecting the doctor to advise getting rid of the animal. Mrs A was dreading Caroline's reaction to this since she was deeply attached to the cat.

After listening carefully to their story Caroline's doctor skin-tested her with house dust mite, grass pollen, cat fur, and several other allergens. As he suspected, Caroline was not affected by hay fever—her grass-pollen skin test was completely negative. To Mrs A's surprise this was also the case for the cat-fur test. However, Caroline showed a large reaction to house dust mite. To her relief her doctor was able to reassure Caroline that her symptoms were not due to her pet but to the house dust mite and was able to advise Mrs A on measures which, in addition to Caroline's treatment, would help to minimize her daughter's symptoms. (These are described fully in Chapter 11.)

A negative response to skin-prick testing usually indicates that the patient is not sensitive to that allergen. However, in some elderly people the skin may not be capable of reacting and skin-testing is of no value. Negative reactions may also occur if the patient is taking antihistamines, which block the effect of histamine, one of the major chemicals released by mast cells. For this reason, these tablets should not be taken for several days before the skin test. For poorly understood reasons, skin-prick testing with food allergens is less reliable than testing with inhaled allergens, such as dust and pollens, and false negative reactions often occur. Food allergy is better diagnosed using exclusion diets, which are explained in Chapter 10.

Patch-testing Patch-testing differs from skin-prick testing and intradermal testing in that it is not used to investigate allergies affecting the nose or lungs but is a specialized test used to identify substances which give rise to allergic skin disease such as dermatitis. Patch-testing is popular because, if the allergen can be discovered, further contact can be avoided, leading to a cure. Patch-testing is time consuming and can occasionally give misleading results. An area of normal skin not affected by dermatitis is usually tested, often the back or occasionally the upper arm. When the doctor applies the tests he or she will ask the patient to sit in a relaxed position, with arms at sides, to avoid stretching or wrinkling of the skin of the back. A series of small discs, each impregnated with a small amount of the substance to be tested, will then be applied to the patient's back. Once the rows of tests have been applied, hypo-allergenic tape will be fastened over them to keep the patches secure. The patches are kept in place for 48 hours and must be kept dry. After the tapes have been removed it is all right to wet the area, but the patient will be asked not to scrub or rub the test sites! The tests are normally read the day after the tapes have been removed. The reaction can be delayed and the patient may need to see the doctor again a few days later. The tape itself may cause slight itching but if *severe* itching or discomfort occurs the patient should remove the patch immediately. Responses to patch-testing help the doctor decide what substances may be affecting the patient's skin. Interpretation of the tests can be difficult and the patient should be guided by the doctor's assessment of which positive reactions are relevant. It is obvious that patch-testing with strong irritants such as undiluted petrol would produce irritant reactions. However, many other materials, for example turpentine or chrome salts, can be irritating under patch-test conditions in some patients. These irritant reactions can be confused with allergy.

Patch-testing does not mimic normal exposure and this can cause problems. Clothes often cause allergy but frequently as a result of associated sweating and friction, particularly around the armpits— factors that cannot be reproduced in a patch test. A similar problem exists in investigating the effects of medicinal creams and ointments which are usually applied to inflamed skin. Since normal skin is liable to be more 'resistant' to penetration by the medication, the patch test often has to carried out using much higher concentrations than would normally be used in treatment.

Determining the relevance of positive patch tests to the patient's skin rash often involves a considerable amount of detective work on the part of the patient and the doctor. For example, positive patch tests to chrome in patients suffering from dermatitis caused by cement were considered puzzling but irrelevant until it was eventually shown that cement contains minute amounts of chrome salts.

Breathing tests

It is vital to test the function of the lungs if a patient has asthma. These tests include:

(1)   those which give information about the severity of asthma; and

(2)   those which investigate the allergens aggravating the disease.

Air flows quickly and easily through normal airways but if the airways are narrowed by asthma, airflow is slowed. The degree of slowing, described as airflow limitation, is measured by tests of lung function. The simplest instrument used to assess airflow limitation is a peak flow meter which measures the maximum rate at which air can be blown out of the lungs. The worse the asthma, the slower the rate of airflow. Indeed, in severe asthma airflow may be so slow that the value is unrecordable. Other factors which affect the peak flow rate are:

(1) height—chest size;

(2) age—the lungs are largest during teenage years, and steadily decline thereafter; and

(3) sex—men have larger lungs than women.

Normal values are shown in Fig. 27.

A more sophisticated device known as a Vitalograph is often used to measure lung function. In addition to recording the amount of

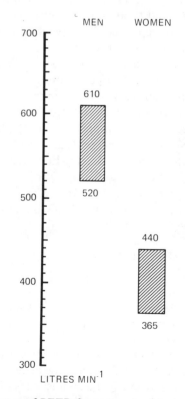

Fig. 27 Normal ranges of PEFR for a person of average height between the ages of 20 and 60 years.

airflow limitation of the lungs, the volume can also be measured. The patient blows forcibly into the Vitalograph and continues blowing until all air is expelled from the lungs. The volume of air expelled in the first second is known as the forced expiratory volume in one second (FEV_1) and is dependent on the same factors which influence the peak flow rate. The total volume of air expelled is known as the forced vital capacity (FVC). Diseases such as asthma and chronic bronchitis cause a reduction in the FEV_1 but relatively little reduction in the FVC. The ratio of the FEV_1 to the FVC is normally about 80 per cent. In asthma the ratio is reduced to below 70 per cent and in very severe disease to as low as 25 per cent.

Fig. 28   Using a Vitalograph.

As mentioned in Chapter 6, breathing problems in asthmatics are variable and they may attend the clinic at a time when their airways are not narrowed and their FEV_1 and FVC appear normal. In this case, the next step is to monitor the airways over a period of time, usually two or three weeks, by giving the patient a portable plastic peak flow meter, such as a Mini-Wright peak flow meter, to blow into several times a day, recording the results on a diary card. In this way changes in airflow can be detected. Some people develop asthma on exposure to allergens at work, for example solderers,

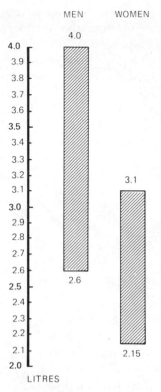

Fig. 29   Normal ranges of FEV_1 for a person of average height between
the ages of 20 and 60 years.

bakers, grain handlers, and many other employees in a variety of occupations. Asthma occurring at work would not be detected in the clinic but would show up in peak flows recorded while at work. Similarly, some people may only develop asthma on exposure to grass pollen or when exposed to high levels of dust, and in these situations peak flow monitoring is an excellent method of confirming a diagnosis of asthma.

In monitoring breathing with a portable peak flow meter, as with all the above tests, the patient's full co-operation is necessary if the doctor is to obtain meaningful and helpful results, since most tests rely on air being forced out of the lungs with the greatest possible effort.

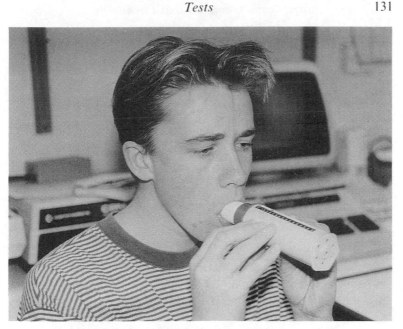

Fig. 30 Use of Mini-Wright peak flow meter.

Hyper-responsiveness The diagnosis of asthma is usually made from symptoms and the results of breathing tests. If peak flow or FEV$_1$ are reduced and return to normal after inhaling a bronchodilator aerosol, the diagnosis of asthma is confirmed. However, in some cases symptoms may not be typical of asthma, for example a cough with no wheezing, or shortness of breath and normal breathing tests. Under these circumstances other tests are needed to prove the diagnosis of asthma.

The airways of asthmatics' lungs are more sensitive to inhaled chemicals than those of normal individuals. This 'twitchyness' of the airways is known as hyper-responsiveness and is present to some degree in all asthmatics, as explained in Chapter 6. Hyper-responsiveness is measured by bronchial provocation testing in which the patient is asked to inhale aerosols of a series of solutions of histamine or methacholine which gradually increase in strength. The strength of solution which can be tolerated before a mild attack of breathlessness or wheezing is induced is an indication of the responsiveness of the airways. The weaker the solution provoking a reaction, the more sensitive the patient and, in general, the greater

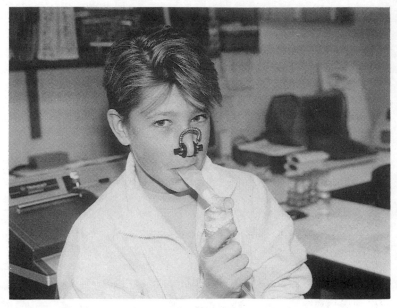

Fig. 31   Child and bronchial provocation test (see text).

the need for treatment. The twitchyness of the airways in the lungs is influenced by exposure to allergens and could be normal in grass-pollen-sensitive asthmatics tested in the winter time.

Once the diagnosis of asthma has been made by breathing tests, it is often necessary to decide whether or not the symptoms are a result of an allergy and, if so, to identify the problem allergen. Usually this is done on the basis of symptoms and skin or blood tests, but sometimes this is insufficient to reach a diagnosis. Under these circumstances provocation tests with allergen are necessary. This is particularly so in cases of occupational asthma where no blood or skin tests are available. These tests involve the inhalation of minute amounts of allergen in a similar way to provocation tests with histamine or methacholine. Once again, gradually increasing strengths of allergen solution are inhaled and lung function closely monitored until a mild attack of asthma occurs. Potentially this test is more hazardous than inhalation of histamine or methacholine, and all these tests must be carried out only in well-equipped hospitals and clinics with specially trained staff to ensure that risks to the patient are kept to a minimum.

Exercise testing Exercise provokes wheezing and shortness of breath in most asthmatics. Hence, most untreated asthmatics rarely participate in sporting activities. The inhalation of cold, dry air aggravates the twitchy lungs of asthmatics causing lung function to fall. This can be used to diagnose asthma, particularly in children who are happy to run around the doctor's surgery or hospital grounds. Peak flow is measured before and after exercise and a fall diagnoses asthma.

Blood tests for allergy

Increasingly, blood tests are used for allergy diagnosis. The tiny amounts of antibodies causing allergy can be accurately identified in specialist laboratories. Only a small sample of blood is needed.

PRIST PRIST stands for 'paper radio-immunosorbent test' and measures the total level of IgE in the blood, in other words the total amount of antibodies causing allergy. It is used as a screening test. If the total IgE is below a certain level, usually 100 international units per millilitre of blood, allergy is most unlikely. *Very* high levels of IgE are found in:

(1)   worm infestations;

(2)   some types of cancers;

(3)   allergic bronchopulmonary aspergillosis; and

(4)   some types of eczema.

The test is performed on a small sample of blood in a laboratory. It is very complex and involves the use of radioactivity. Full details can be found in Appendix B.

RAST RAST stands for 'radio-allergosorbent test' and measures the amount of IgE antibody against individual allergens, for example grass pollen or house dust mite. The results are graded from 1–4, where 4 is the most positive. (See Appendix B for details of the test procedure.) However, the results of RAST, like skin-testing, do not give a definitive answer and must be related to the patient's history. RAST has the following advantages over skin-testing:

(1)   it is not influenced by any tablets;

(2)   it can be performed when there is widespread skin disease;

(3)   it is completely safe;

(4)   it is very useful in research since blood samples can be stored for use in the future; and

(5)   it is more likely to be positive in cases of food allergy.

The disadvantages are:

(1)   the result is not immediately available, unlike skin-prick testing;

(2)   the level of specific IgE in the blood can vary—some hay-fever patients have negative tests before the hay-fever season, which only become positive following exposure to grass pollen during the summer.

MAST   A system similar to the RAST system has been developed recently. This has the benefit of using luminescence rather than radioactivity as the marker of a positive test. It is simple to perform and does not need the facilities of a specialist laboratory. As many as 35 different allergens can be detected in one test. Again, full details can be found in Appendix B.

Eosinophil counts   One of the simplest blood tests related to allergy is to count a particular type of cell in the blood. Eosinophils are white blood cells which stain with the red dye eosin (hence the name) and can easily be seen under a microscope. Usually these cells make up 3–5 per cent of the total numbers of white blood cells.

   Allergic reactions cause more eosinophils to be made and also attract them from the bloodstream to the site of the allergic reaction. Two factors affect the extent to which eosinophil levels are raised:

(1)   the size of the organ where the allergic reaction is occurring; and

(2)   the severity of the inflammation there.

Since the lungs are so much larger than the nose, increased blood levels of eosinophils occur more often in asthmatic patients than in those with rhinitis. Because eosinophils are related to inflammation they reflect better the severity of a disease than the degree of allergy involved. In allergic diseases eosinophils increase to between 500 and 1500 cells per millilitre of blood and in moderately bad asthma to between 1000 and 2000 cells. Very high numbers suggest other problems, including:

(1)   worm infestation;

(2)   skin disease;

(3)   drug allergy; and

(4)   Hodgkin's disease (a type of cancer).

Interestingly, steroid tablets cause a reduction in eosinophils in the blood, and it is important that the doctor knows if the patient is taking steroid tablets when the blood test is taken.

Nasal tests

Nasal provocation testing   It is generally considered that what is needed to diagnose seasonal allergic rhinitis due to grass pollen (hay fever) is a strong story from the sufferer supported by a positive skin-prick test to grass pollen. In other cases of rhinitis the answer may be less clear-cut, especially in the case of perennial allergens, such as house dust mite and cat fur, where exposure to dust in the home may cause symptoms due to the house dust mite, to cat fur and dander shed into the dust, or both. The problem of identifying which allergen is mainly responsible for nasal symptoms is made even more difficult if the sufferer has positive skin-prick tests to both suspected allergens. It is important to decide which allergen is responsible, and a nasal provocation test may be suggested. This test involves applying a small amount of the suspected allergen to the lining of the nose and recording any reaction which occurs. There are several ways in which the test substance may be applied, which include:

(1)   as a spray;

(2)   impregnated onto a paper disc; or

(3)   on a cotton bud.

The response to nasal provocation testing is usually immediate and resembles classical hay fever with one or more of the following symptoms:

● sneezing

● nasal blockage

● nasal discharge

● itching

There are a number of ways in which the symptoms provoked may be assessed in the clinic. One of the simplest is for the doctor to inspect the lining of the patient's nose to observe changes in the degree of redness. It is also possible to count the number of sneezes which occur in the 15 minutes following application of the allergen and, if the patient sits with head forwards, measure the secretions produced. When used in combination these two measures can provide a 'sneeze–drip score' giving some indication of the intensity of the patient's response. Such scores are often compared with the results of testing with ordinary saline solution which acts as a control test and allows the doctor to differentiate between a true allergic reaction and a purely irritant one. More accurate records of changes in the nose can be obtained using instruments which measure nasal blockage during nasal provocation testing, such as:

(1)   a rhinomanometer (which measures flow and pressure in the nose); or

(2)   a simple nasal inspiratory flow meter (SNIF meter).

The SNIF meter closely resembles the Mini-Wright portable peak flow meter used to assess asthma, but is fitted with a mask to enable nasal flow to be measured. In contrast to the lungs, where peak flow rates of exhalation are measured, the nasal meter measures the fastest rate of airflow if the patient sniffs in forcibly.

A more complex instrument for measurement of nasal blockage is the rhinomanometer, which measures pressure and flow in the nose either via a mask placed over the patient's face or via a small flexible tube placed just inside the mouth. Recordings are quick and painless and can be carried out either while breathing quietly (active rhino-manometry) or while the breath is being held (passive rhinomano-metry). Whatever type of provocation test is recommended, the results can be greatly affected by some of the medications which the patient may be taking, particularly antihistamines or steroid sprays. It is very important that the patient knows how long before the test to stop taking the medication.

Nasal smears   In the section relating to blood tests a simple investigation involving the counting of eosinophils, a type of white blood cell, was discussed. These cells are attracted from the blood-stream to the site of an allergic reaction, for example the lining of the nose. It is possible to count the numbers of eosinophils on the surface of the nasal mucosa by taking a smear. Such smears can be obtained by using a tighly wound cotton swab (cotton bud). The swab is inserted two or three times into each nostril and the lining of the nose scraped with a firm, rolling movement. The test is not painful but may be slightly uncomfortable since smears need to be taken from as far back in the nose as possible. Smears taken from just inside the entrance to the nose only consist of a watery discharge rather than a good sample of nasal secretion which is likely to contain cells.

Secretions recovered on the swab are spread gently onto a microscope slide and sent to the laboratory to be stained and to have the cells counted under the microscope. Reading the slides is a skilled and time-consuming job, not least because the cells often appear in clumps and the entire smear must be studied on every occasion. If more than 10 per cent of the cells seen are eosinophils (eosinophilia), this indicates that the smear is from a person suffering from allergic rhinitis.

A nasal smear can help to differentiate between patients with allergic rhinitis and those whose symptoms really are due to 'recurrent colds', as 80 per cent of patients with allergic rhinitis show eosinophilia as opposed to only 8 per cent of patients suffering from virus infections. However, it is not possible to get any clues as to the cause of allergic nasal symptoms from the results of a nasal smear.

Tests for food allergy

Many people know from experience that particular foods cause them to react badly, and they therefore take steps to avoid these foods. Frequently, however, the picture is less clear. Some reactions occur after a delay of several hours, or even days; or symptoms may be chronic, occurring day after day. In these people the problem food, if any, is more difficult to identify but can usually be found by a combination of dietary manipulation and food challenge. The first step is usually to restrict the diet drastically, often to one which consists simply of lamb, pears, rice and/or potatoes, and bottled water. More details of different types of diets used can be found in Chapter 10, which deals with food avoidance. A restrictive diet should not be undertaken without seeking the advice of your doctor and/or dietician, since these diets normally need to continue for 10–21 days before further investigations can begin and during this time *nothing* other than the prescribed diet can be eaten. It is particularly important to discuss dieting with your doctor when the patient is a child, or suffers from any chronic illness, or is taking any medication.

If, after three weeks on a restricted diet, symptoms have not improved, it is extremely unlikely that food is the cause of the problem and further dieting is pointless. However, if the diet has resulted in some improvement 'test' foods can begin to be reintroduced one by one. This is a long, slow process but is the only way to identify both those foods which cause problems and those foods which are well tolerated. The rate at which foods can be reintroduced into the diet is different for each individual, since some people react within a few hours and others not until the following day. Advice of food allergists who suggest introduction of several 'test' foods each day should be treated with caution, as in this case it is impossible to associate delayed effects with the correct food.

Because diseases like asthma and rhinitis are variable in their own right, an attack of breathlessness or sneezing after testing a particular food could easily be coincidental. For this reason, unless a dramatic reaction has occurred, doctors usually advise repeating the test after the symptoms have been allowed to subside. If the same reaction is provoked, especially if this occurs at the same time after eating the test food, it is very probable that the food is the true cause of the problem.

If a food is suspected of being responsible for an acute or life-threatening reaction, such as anaphylaxis or severe asthma, it is especially important to identify the food so that it can be scrupulously avoided in future. However, challenge testing with such foods is obviously associated with some risk and, if absolutely necessary, is only attempted under strict medical supervision with very tiny amounts being reintroduced at first, gradually building up to a 'normal' portion.

Often when food allergy is suspected either the patient or doctor may have a preconceived idea about the cause. In a test situation the patient may, when faced with the food, anticipate a reaction and subconsciously influence the test result. For this reason test foods are sometimes given in the form of 'double-blind' or 'single-blind' challenges carried out in a hospital. In a double-blind challenge two tests are usually performed. On one occasion the real food being tested is given to the patient and on another occasion it is substituted by a 'dummy' food which acts as a control test. Neither patient nor doctor knows which test is which since test feeds are prepared in advance by a third person. Sometimes test foods are given as coded capsules which are indistinguishable from each other, or as a soup to disguise the taste of the food. On rare occasions foods which cannot be easily disguised are given in liquid form through a small tube directly into the stomach.

Single-blind testing is carried out in exactly the same way, except that in this case the doctor, but not the patient, is aware of the nature of each test.

Testing for food allergy can be slow and tedious and requires the co-operation of a very 'patient' patient, but it can be valuable in proving that it really is necessary and beneficial to follow a special diet in the long term.

X-rays

The X-ray is one of the best known tests used in the investigation of chest problems. In patients with asthma the chest X-ray is usually normal and is taken to exclude other diseases rather than to confirm the presence of asthma. The results of the X-ray do not help to identify any allergic component of the disease. In contrast, an X-ray of the sinuses in the face can be useful since 40 per cent of patients with perennial rhinitis also have sinus problems, often needing antibiotic treatment.

Private tests

A number of tests for allergies to common environmental allergens, including those in foods, are offered by several private allergy clinics. These include tests on:

(1)   blood cells

(2)   blood spots on tissue; and

(3)   locks of hair.

These allergy tests are expensive. They are not reliable. *Which?* magazine has investigated private allergy tests, comparing their results with those of skin and RAST testing as described in this chapter. Important allergens, for example allergy to fish in patients known to react to this food, were not reliably identified. In addition, different results were obtained when two samples from the same patient were sent to the same laboratory on different occasions. *Which?* considered that the advice sent back with the tests was often dubious and at worst risky. The magazine pointed out the importance of appropriate controls over these private allergy clinics, which need to be run by suitably qualified and trained allergists.

10

AVOIDING ALLERGENS

Obviously the most satisfactory solution to the problem of allergy is to remove the offending allergens from the patient's environment. Although it is impossible to eradicate many environmental allergens, avoidance measures can be helpful. Such measures involve alterations to lifestyle, diet, occupation, and leisure activities, and the extent to which a sufferer's family, social, and domestic circumstances are affected by such changes must be balanced against the degree of disability and severity of symptoms. Many families find the idea of parting with a cat or dog intolerable and consider allergic symptoms an acceptable alternative; similarly, those with important business or social commitments may find it impossible to follow a rigid exclusion diet, and such radical avoidance measures may cause almost as much distress as the allergic symptoms themselves. A sensible, balanced approach to treatment must be considered, in which allergen avoidance can play an important role.

The house dust mite

Dermatophagoides pteronyssinus and *Dermatophagoides farinae*, the commonest house dust mites, are present in varying numbers in virtually every home. The mites thrive in damp, humid conditions, and it is therefore not surprising that they are one of the major causes of allergic disease in the United Kingdom. The use of domestic air conditioning and dehumidifiers can be of great value in minimizing the conditions which favour mite reproduction. Mites occur in the greatest numbers in bedding, particularly in mattresses, and sufferers will often notice that symptoms are worst when the bedding is disturbed, and allergenic material becomes airborne. Indeed, there may be up to 4000 mites in each gram of surface dust from mattresses.

Unfortunately, even larger numbers of mites can be found in soft toys and teddy bears, and a child cuddling one for comfort during an asthma attack may unknowingly be making matters worse. The major source of mite allergen is the faecal pellet, and measures to limit the skin scales available for mite ingestion and to minimize the conditions for mite reproduction are valuable in controlling exposure.

The following measures can lead to reduction in mite numbers and have been shown to reduce symptoms in controlled clinical trials with children.

The bedroom

1. Have a bed with a plain wooden base, not a divan.

2. If possible buy a new mattress, and enclose it immediately in a washable plastic cover: nylon covers are of no value. If an old mattress is to be covered, ensure it is thoroughly aired to avoid condensation before enclosing it. A cotton under-blanket will aid comfort.

3. Remove feather pillows and eiderdowns, replacing them with washable synthetic ones.

Fig. 32 A haven for the house dust mite.

4. Replace woollen blankets with synthetic or cotton ones or with a washable duvet.

5. Change and wash pillow cases, sheets, blankets, duvet covers, and under-blanket every week. If possible wash the bed-linen at a water temperature exceeding 130 °F which will destroy any mites present. Sponge and dry plastic mattress covers each time the bed is stripped.

6. Remove carpets and cover the floor with wall-to-wall linoleum or vinyl flooring, with small washable rugs if required.

7. Have light, washable curtains or roller blinds.

8. Do not allow animals into the bedroom. Mite counts can be raised in areas frequented by domestic cats and dogs, particularly where the animal sleeps.

9. Do not use a lower bunk if the one above is occupied: movement during sleep will cause a constant fall of allergenic particles from the upper bunk onto the sleeper below.

10. Machine wash soft toys regularly, particularly those taken to bed! Soft toys for allergic children should be purchased with care to ensure that they are fully washable.

Hospital wards, where most of the features mentioned in this list apply, have extremely low levels of mite infestation.

Housework

The good news for sufferers from house dust mite or house dust allergy is that a non-allergic member of the family should be asked to take on the housework that can cause allergens to become airborne. A few sensible precautions can be taken if this is impossible.

1. Use a vacuum cleaner with disposable paper bags or the non-porous bags designed to contain allergenic particles; never empty one out.

2. Do not brush or shake carpets and rugs; never beat any upholstered furniture. Upholstered chairs and sofas contain high mite counts but leather, vinyl, or washable covers are of some value in lowering mite levels.

3.   Always use a damp cloth or mop for dusting or cleaning floors, and steam clean carpets if practicable.

Choice of homes

In so far as this is possible, avoid old or damp accommodation and property near overground or underground rivers or canals. Avoid ducted air central heating, though other types of central heating are helpful in reducing humidity. Do not purchase property in need of renovation or structural repairs: such works greatly increase dust levels and cause months of disturbance in the atmosphere of the home. Avoid properties where any signs of mould are evident, remembering that exposed north-facing walls can be very damp. Frequent air changes are necessary both to reduce humidity and levels of airborn allergens, so it is essential that rooms are aired regularly and a good through-draft is achieved. Fixed glazing should be avoided, and double-glazed units, which help to retain heat and humidity, must be opened regularly in spite of the increased heating costs!

There are many different species of mites, and sufferers from allergy to the house dust mite may suffer cross-reactivity with other species. Care should be taken to minimize the conditions under which the different species may occur in the home.

Storage mites, for example, may enter an encysted state in grain and emerge unscathed through processing to contaminate flour. Cereal foods kept for too long may develop a mite population, as may the crusts of cheeses, such as Stilton, which are warm and moist and provide an excellent environment for mite growth. Mites thrive at a warm temperature and in materials with a water content of 14–16 per cent so cool, dry storage is recommended. Cereal and flaked foods, for pet animals and fish, may also be infested and should similarly be carefully stored. Old, damp accommodation, which as already mentioned provides optimal conditions for house dust mite infestation, may encourage the growth of moulds, which themselves provide a useful food supply for the storage mite species. Treatment of moulds, disturbing both mites and mould spores, should not be undertaken by, or in the presence of, a mite-sensitive individual, and the room should be aired thoroughly and the atmosphere allowed to settle before the room is used.

Clearly these measures are temporary and must be painstakingly sustained, but what can be done to cure the problem of exposure to the house dust mite? Currently the use of acaricides (chemicals designed to kill the mite, such as IGR-5, a methaprine acaricide) is under careful study, but there are as yet no validated results to prove the efficacy of these agents in the domestic setting, and the possibility that such chemicals will prove toxic to man needs careful evaluation.

Investigators are also studying the role of a natamycin fungicide (Timosyl) for use in house dust mite infestation. Various fungi, particularly *Aspergillus penicilloides*, contribute to the diet of mites, though whether this is to aid the digestion of skin scales or for essential amino acids not found in the human epithelium is uncertain. In the laboratory it has been shown that removal of fungi from the mite's diet reduces the reproductive rate of the mite, so reducing its numbers. Whether this reduction will be reflected in significant relief of symptoms in allergy sufferers will not be known until the double-blind studies presently underway are complete.

Non-porous plastic mattress covers can be extremely uncomfortable and can give rise to excessive night-time perspiration. A study of a new cotton-backed polyurethane with a pore size of 0.5 µm (five-hundredths of a millimetre) is presently being conducted. The aim is both to starve the mites of human epithelial scales and to prevent the 20–30 µm faecal pellets of the mite from passing through the cover. If successful, mattresses could be made using this material, hopefully preventing establishment of a mite population. A vacuum-cleaner bag of the same material is also being examined, the aim being to reduce the amount of small dust particles which escape through the walls of conventional paper bags.

Holidays

Again, avoid old or damp accommodation and property close to water. Avoid accommodation which has not been used during the winter months as the mites will have proliferated while undisturbed. Take your own pillow and mattress cover with you, and if you use sleeping bags wash them and store them in a dry place when not in use. As mites do not live at altitudes greater than 2000 metres above sea-level, a holiday in the Alps is highly recommended!

Pollens

The range of grass, tree, and shrub pollens to which individuals may be allergic is extensive, each having a slightly different season (see Chapter 3).

It is possible to reduce exposure to pollens to some extent by:

(1)   not sleeping with bedroom windows open;

(2)   keeping car windows closed when driving;

(3)   avoiding going out in the early evening when pollen grains, which have risen up into the atmosphere during the day, fall as the air begins to cool;

(4)   getting someone else to mow the lawn or cut grass;

(5)   not keeping pollinating plants in the home;

(6)   taking holidays at the coast, where sea breezes keep pollen grains inland; and

(7)   wearing glasses which may help to prevent pollen grains from entering the eyes, thus reducing symptoms of allergic conjunctivitis.

However, these measures are rarely, if ever, sufficient to stop the symptoms of seasonal rhinitis or conjunctivitis.

Animals

Sensitivity to animal allergens is well recognized and can be due to skin scales (dander) or substances (proteins) present in urine or saliva. Such sensitivity can cause minor inconvenience or may be an important occupational problem.

The solution in the domestic environment is to remove the pet and thoroughly clean the home to remove the accumulated allergen. If this is not possible, a sensible approach must be taken to minimize exposure, and the following measures should be of help.

1.   Never allow animals into the bedroom.

2.   Set aside areas in the home where pets are not allowed.

3.   Rooms where pets are allowed should have washable flooring which will minimize the accumulation of allergen particles.

4.   Pets should sleep in a shed or outhouse. Their bedding should be

changed or washed regularly. Washable plastic baskets are preferable to cane, which harbours danders and mites.

5. Grooming and clipping should not be done in the house or by anyone allergic to the animal in question.

6. Discourage animals from licking people; wash immediately if animal saliva comes into contact with the skin.

7. Do not allow visitors to bring animals into your home. Where visits to homes with animals are unavoidable, ask the owner if it is possible for the animal to be kept in another room to avoid direct contact, though contact with the danders present in the environment may still give rise to symptoms, particularly in cat allergy. This can be a very sensitive issue and needs careful handling!

8. Discourage animal-allergic children from touching or handling animals, and encourage them to wash immediately after contact when this is unavoidable.

9. Do not buy or replace pets in future.

Individuals who are atopic, particularly if they are allergic to animals, should avoid employment which involves contact. Farm workers who become allergic may have no alternative but to leave employment to avoid contact with animals. Allergic symptoms in laboratory workers commonly occur from contact with small mammals such as rats, mice, rabbits, and guinea-pigs. It is essential that:

(1) animal houses are well ventilated and dust levels reduced to a minimum; and

(2) proper protective clothing is worn when cages are cleaned or animals are handled.

Occupational exposure

There are few circumstances in which the maxim 'prevention is better than cure' is more appropriate. More and more people are developing allergy or sensitivity to dust, gases, vapours, and fumes encountered at work. The first principle of occupational hygiene is to reduce exposure to harmful materials to a minimum, and regulations concerning the levels allowed at work are published annually by the Health and Safety Executive. The latest Guidance Note (EH 40/86) gives advice on limits to which exposure to airborne substances hazardous to health should be controlled in the work place. The harmful effects of materials encountered at work are related both to the amount present in the air and the time during which contact occurs. These two factors are combined in the time weighted average (TWA) which is an expression of exposure over an eight-hour shift. Recommended limits of exposure have been set by the Health and Safety Executive which represents good industrial practices, plant design, and engineering controls. Some examples of recommended control limits, for industrial materials, expressed as TWA in milligrams per cubic metre of air, which should not normally be exceeded are:

Isocyanate	0.02
Formaldehyde	2.5
Phthalic anhydride	6.0

People can, however, become allergic or sensitized after exposure to quantities of the offending agent well below the TWA. Ideally, all processes which liberate harmful dust, gas, vapour, or fumes should be totally enclosed, with safe extraction, well away from the work site. Great care must be taken to ensure that extracted materials are safely disposed of, since asthma due to sensitization to isocyanates has occurred in employees whose job did not involve their use. Exposure resulted from air containing isocyanates being sucked into the ventilation system of an office block. The contamination of the atmosphere was the result of discharge of isocyanate fumes from a nearby factory.

Protective clothing and masks

It may not be feasible to enclose all hazardous processes or provide adequate extraction. Under these circumstances it is essential that people are provided with protective clothing and masks.

Hypo-allergenic cosmetics

Hypo-allergenic cosmetics, as their name suggests, contain the minimum amounts of potentially allergenic constituents consistent with production of safe but attractive cosmetics. It may not be assumed that these products are completely allergen-free. Both carefully selected natural and synthetic ingredients are used, some manufacturers preferring the latter as they are more easily standardized. The lowest possible amounts of preservatives and anti-oxidants are used, and perfume omitted. However the use of chemical additives is important for the presentation, shelf-life, and safety of cosmetics. Failure to prevent contamination of cosmetics by micro-organisms may itself give rise to symptoms of skin inflammation and conjunctivitis. It is advisable to discard old cosmetics which may be the cause of unpleasant symptoms before trying to elucidate an allergic causation.

Preservatives are used to inhibit the effects of bacteria, yeasts, moulds, and a wide variety of other micro-organisms. Those widely used in hypo-allergenic cosmetics, such as the esters of parahydroxybenzoic acid, have a low toxicity and a low incidence of contact sensitization.

Anti-oxidants are used to prevent oxidative deterioration and to prevent fats and oils from turning rancid. Butylhydroxyanisole (BHA) or butylhydroxytoluene anti-oxidants are used by some manufacturers, while others use a natural anti-oxidant, tocopherol, and package products in sealed tubes, or bottles sealed under a nitrogen atmosphere.

Colouring agents are essential to the manufacture of cosmetics, and about 100 are at present authorized for use in cosmetic products. Although these may be considered non-toxic, a great number of them may be potentially allergenic. In particular, reactions may be found to azo-colourants, eosin and its derivatives, chromium greens, cobalt blues, and various organic colourants or pearly substances. Where colourants are unnecessary, as in skin-care products, these are omitted.

Perfumes are complex chemical products and may contain over 100 different constituents, many of which may be allergenic. As their only function in the manufacture of cosmetics is to detract from the smell of certain raw materials, they are omitted in hypo-allergenic cosmetics.

Lanolin is a natural product from the sebaceous glands of sheep. It closely resembles sebum and is a good substance for use on human skin. However, before it is suitable for use in cosmetics, it must be transformed from its original 'wool grease' form to lanolin, and this may involve the use of potentially allergenic chemicals such as formaldehyde. Only high-grade lanolins should be used in the manufacture of cosmetics, with carefully regulated purification processing. Agricultural chemicals and insecticides may be responsible for the allergenicity of a particular batch of lanolin, while other batches, even from the same source, may be perfectly safe to use. Allergy-screened cosmetics containing lanolin are worth trying, even for those who have suffered discomfort from using lanolin in the past.

For many, common sense and preference dictate the use of the purest possible products on the skin and around the eyes. Reputable cosmetic houses perform dermatological testing on all products prior to marketing and guarantee strict standardization. Adverse reactions to cosmetics may often be irritant rather than allergic and a physician's assessment is necessary to determine the nature of the reaction and to advise appropriate avoidance measures. For example, primary irritation by a detergent or solvent may cause damage to the skin and promote the onset of a secondary allergic reaction to a cosmetic colouring agent.

Powdered skin preparations are less likely to cause problems than their cream or liquid forms.

The eyelid and surrounding tissues should be clean before application of eye make-up and excessive rubbing during cleansing should be avoided. The omission of perfumes from eye make-up products generally has removed one recognized source of sensitivity.

Lipstick may be responsible for dermatitis of the lips (cheilitis), caused by perfumes or lanolin. Benzoate preservatives used in lipsticks may cause more generalized reaction in other parts of the body due to their unavoidable ingestion. Toothpastes containing benzoate preservatives should similarly be avoided in cases of known or suspected sensitivity. Patients following colouring-, preservative-, and salicylate-free diets should be careful to check whether they are accidentally ingesting benzoates in this form.

Nail varnishes are responsible for many cosmetic-induced allergic reactions. They may contain either a toluene sulphonamide or formaldehyde resin base. Dermatitis develops away from the actual site of application, commonly on the eyelids, lips, chin, neck, or armpits, or other areas regularly in contact with the nails. Polyester resin may be used as an alternative base.

Many of the major cosmetic houses specializing in hypo-allergenic cosmetics run a physician's information and advice service, and provide full information of constituents of their products upon application.

For reasons of patent security, general manufacturers may be unable to provide such full information, but if in doubt about the suitability of a particular product or range of products it is worth writing and specifying the agents to be avoided, and asking which products may then be recommended for use.

Laboratoires RoC (UK) Ltd, of 13 Grosvenor Crescent, London SW1X 7EE, offer a very comprehensive and helpful advisory service.

Almay Products, of 225 Bath Road, Slough, Berkshire, have staff available to discuss specific problems regarding reactions and offer assistance to individual customers seeking to avoid specific ingredients.

Boots of Nottingham produce a range of allergy-screened products and will answer individual queries about their products, which should be addressed to Consumer Product Development, Boots Company PLC, 1 Thane Road, Nottingham NG2 3AA.

Clinique run a Physician Service Centre at 54 Grosvenor Street, London W1X 9FH, providing product information, formularies, and patch-test materials.

Dietary management of food intolerance

Symptoms of food intolerance can often be avoided by elimination of the offending food. However, it is important to ensure that everyone's diet is nutritionally adequate, providing sufficient protein, energy, vitamins, and minerals. This is especially important in young children whose growth rate can be impaired by poorly designed, low-energy, elimination diets. If very restricted diets are necessary, daily supplements are essential. Do not adopt a very restrictive diet without professional help. Ideally, physician and dietician should work together planning food avoidance diets.

Elimination diets

1. *Exclusion diets* are used if one or two foods regularly cause symptoms. These foods are then excluded from the diet.

2. *Oligo-allergenic* (low allergen) diets are used to avoid some of the foods most commonly associated with food allergy, for example diets free of dairy produce.

3. *Strict elimination* diets allow a very restricted number of foods, for example one meat, one fruit, one vegetable, and one starch source. The choice of foods is based on those least likely to cause allergic reactions, such as lamb or chicken, rhubarb or pears, carrots or sprouts, and rice or potato. A fat source, such as a milk-free margarine or a specific vegetable oil, and sugar may also be allowed. These strict elimination diets are often nutritionally inadequate and need to be supplemented with vitamins and minerals. Energy and protein supplements are also given in young children and in adults if these diets are to be used for more than a few weeks.

4. *Elemental* (chemical) diets are sometimes used as a last resort to enable a diagnosis of food intolerance to be made or rejected with absolute certainty. Such diets are based on purified food in powder form and are rather unpalatable. The powder is made into a drink using purified water.

Milk-free diets

Such diets involve exluding all types of milk (fresh, dried, and tinned) and milk products, including yoghurt, cheese, butter, and cream. Many manufactured foods contain milk-derived ingredients, such as casein, whey, lactose, lactalbumen, skimmed-milk powder, milk solids, buttermilk, and non-fat milk. Great care should be taken when checking the list of ingredients on manufactured foods to ensure that the food is actually milk-free.

Manufactured foods likely to contain cow's milk and milk products include:

(1) ice-cream, including non-milk fat ice-cream;

(2) milk puddings and powdered desserts;

(3) cakes, puddings, and biscuits

(4) coffee creamers;

(5) baby foods;

(6) margarines and low-fat spreads;

(7) malted bedtime beverages;

(8) breakfast cereals, especially muesli and instant oats;

(9) tinned and packet soups;

(10)   chocolate, fudge, and toffee

Milk-free alternatives do exist to some of these foods. Kosher margarine and some 'vegetarian margarines' from health food shops are suitable. Other kosher foods labelled 'Parev' contain no milk products. Many foods available in any supermarket are suitable for inclusion in a milk-free diet. Do not always assume that all brands of the same food are suitable just because one brand is known to be milk-free (e.g. tinned soups, spaghetti).

Milk is a major source of protein, energy, calcium, vitamins, and minerals, especially in the diet of young children. Milk replacements need careful checking for nutritional adequacy. Many soya-milks are not a suitable nutritional substitute for cow's milk, being lower in energy, calcium, vitamins, and minerals. However, soya-based infant formulas are designed to mimic the nutritional content of cow's milk and should be used for infants and young children. These products can be prescribed for cases of milk intolerance.

Goat's milk is poorly tolerated in up to 50 per cent of cow's milk intolerant patients. It is not subject to the same strict health regulations regarding collection and pasteurization as apply to cow's milk, and often has a high bacterial load. It should always be boiled before use. Nutritional inadequacies make it unsuitable for use in infants under two years old. Goat's milk should be supplemented with vitamin B_{12} and folate, as well as the usual vitamins A, C, and D, when used in infants aged under five years.

If a suitable milk replacement is not taken, it will be necessary to take a daily calcium supplement. One pint of milk substitute should ideally be taken daily in young children, and half a pint daily in older children and adults. Calcium supplements should provide the equivalent of 400 mg elemental calcium daily and not just 400 mg of a calcium salt. Dieticians can suggest a suitable brand and dose.

Some patients are unable to take fresh cow's milk but find that small amounts of heat-treated milk (UHT, evaporated), yoghurt, and cheese are well tolerated. These reactions are individual and need to be found by trial and error.

Milk-intolerant patients often react to other foods, notably eggs. Egg avoidance calls for the exclusion of eggs and all egg-containing foods. The food ingredients albumen and egg lecithin should also be avoided.

Foods likely to contain eggs include:

- cakes, puddings
- quiches and savoury pies
- egg custard
- lemon curd
- Yorkshire pudding
- meringue
- egg noodles

Cereal- and yeast-free diets

A wheat-free diet is difficult as wheat or wheatflour is the main ingredient in many of our basic foodstuffs, such as bread, biscuits, cakes, pasta, and breakfast cereals. Such a diet calls for the avoidance of all foods made from wheat, wheat flour, or wheat-derived products, such as farina, wheatstarch, and hydrolysed vegetable protein.

A yeast-free diet calls for the avoidance of all foods containing yeast, yeast extract, malt, malt extract, malted barley, and hydro-lysed protein. Yeast-containing foods to avoid include: bread; crumpets; teacakes; doughnuts; muffins; yeast-based savoury spreads; foods containing bread and breadcrumbs, including sausages and rissoles; brewers' yeast and vitamin B complex tablets; dried fruits; beers and wines, especially home-brews; malt vinegar; wine vinegar; cider vinegar.

Gluten is a mixture of cereal proteins found in wheat, rye, oats, and barley. Gluten-free diets avoid all these cereal proteins, replac-ing wheat flour with gluten-free flour in a variety of gluten-free manufactured foods (bread, pasta, biscuits, flour mixes) which are available on prescription for patients with coeliac disease and dermatitis herpetiforms (a rare skin disease). The Coeliac Society provides an up-to-date list of gluten-free manufactured foods.

A corn-free diet avoids corn, cornflour, cornstarch, cornmeal, and any vegetable oil or margarine containing corn-oil.

Fish-free diets

Fish intolerance may occur to all types of fish or only specific fish. Shellfish are quite often the culprits. Edible oils found in many processed foods may be of fish origin and should be avoided.

Other elimination diets

Nut-free diets Reactions may occur to all types of nuts or only specific nuts. Bought baked goods should be carefully checked for nuts and marzipan. Ground nut oil should be avoided.

Chocolate-free diet Check for cocoa and chocolate in manufactured foods.

Tea- and coffee-free diets Decaffeinated coffee may be tolerated. Other caffeine-containing beverages should be avoided (e.g. cola drinks).

Soya-free diets Approximately 10 per cent of children with cow's milk intolerance are also intolerant to soya. Foods to avoid include soya milk, soya beans, soya bean curd (tofu), soya bean oil, soya margarine, texturized vegetable protein, and hydrolysed vegetable protein.

Additive-free diets

Food additives, including colouring agents, preservatives, antioxidants, emulsifiers, and flavourings are added to many processed foods. Many people try to avoid additives as they prefer to take a diet based on natural foods. This preference should not be confused with food intolerance.

Food labelling regulations state that most pre-packed and processed foods must include a list of ingredients. Most food additives will be listed, unless they are only active in one of the ingredients and not in the final product. Food additives are usually listed by category name, explaining the function, often followed by the specific name and serial number. Serial numbers with the prefix 'E' have been approved by the EEC. A booklet entitled *Look at the label* lists additives currently permitted in the UK and is available from the Ministry of Agriculture, Fisheries and Foods (MAFF), Publication Unit, Lion House, Willowburn Trading Estate, Alnwick, Northumberland NE66 2PF.

Food additives most likely to cause symptoms

Tartrazine (E102) is the colouring agent most frequently associated with symptoms, possibly due to its widespread use in manufactured foods. Intolerance to more than one additive is common, especially in those sensitive to aspirin (see later).

The use of sodium metabisulphite (E223) has increased dramati-

Table 2

Colourings	E102	Tartrazine
	E107	Yellow 2G
	E110	Sunset yellow
	E122	Carmoisine
	E123	Amaranth
	E124	Ponceau 4R
	E128	Red 2G
	E151	Black PN
	E154	Brown FK
	E155	Brown HT
	E160b	Annatto
Preservatives	E210–219	Benzoic acid and derivatives
	E220	Sulphur dioxide
	E223	Sodium metabisulphite
	E250	Sodium nitrite
Anti-oxidants	E320	BHA
	E321	BHT

cally recently due to its anti-oxidant properties which help to keep fruits and vegetables looking fresh, for example in salad bars. Metabisulphites are used in many other restaurant foods, such as fried potatoes and seafood, and up to 100 milligrams can be eaten in a single meal. Remember that sodium metabisulphite is frequently used in solutions for sterilizing bottles and equipment used in wine-making and home-brewing.

Diets free from most food additives are based on fresh foods and carefully selected manufactured foods. A total additive-free diet is usually unnecessary as many additives have rarely been reported to cause symptoms.

Food labels should be carefully checked. Foods likely to contain suspect additives include:

(1) fruit squash and fizzy coloured drinks;

(2) ice lollies and ice-cream;

(3) batter- and breadcrumb-coated foods, including fish fingers;

(4) smoked fish—kippers, smoked mackerel;

(5) fruit yoghurt;

(6) custard powder and powdered desserts;

(7) boiled sweets;

(8) ham, bacon, and cured meats;

(9) hamburgers, sausages, pâté;

(10) biscuits, cakes;

(11) jam, marmalade;

(12) tinned berry fruits;

(13) tinned vegetables;

(14) breakfast cereals;

(15) potato crisps;

(16) medications, including capsules, tablets, lozenges, and mouthwash, often contain colouring agents.

Salicylate-free diets

Intolerance to aspirin (acetyl salicylic acid) is often associated with sensitivity to foods containing natural salicylates. The response to salicylate-containing foods is often dose-related. Small amounts are well tolerated but larger amounts cause symptoms.

Foods found to cause symptoms include:

(1) fruit (apples, apricots, bananas, berry fruits, cherries, citrus fruits, grapes, nectarines, peaches, plums, rhubarb, and dried fruits);

(2) alcoholic drinks (beer, cider, lager, red and white wine, port, and sherry);

(3) liquorice;

(4) almonds;

(5) peas.

Not all foods listed need to be permanently avoided. It is advisable to find one's own limit by trial and error. Up to two portions of fruit only are usually tolerated.

All hospital doctors and general practitioners have access to National Health Service dieticians. If you suffer from food intolerance make sure you get advice.

Remember, the cure shouldn't be worse than the disease!

11

TREATMENT

Fortunately treatment can be very effective. First, it is essential to find out the cause of the allergic symptoms, and then, wherever possible, avoid the offending allergens. When the cause is a domestic pet, a particular food, something in the environment at work, or a cosmetic, avoidance can be very effective. Unfortunately, at present it is not possible to reduce exposure to the house dust mite sufficiently to cure related allergies, and it is virtually impossible to avoid exposure to pollens or moulds. However, it should be possible totally, or at least largely, to relieve symptoms of allergic diseases provided sufferers are prepared to take the following steps.

Hay fever

Inhalation of the powerful odours of essential oils has been fashionable for the treatment of nasal symptoms for many years. Many sufferers still rely on lozenges and inhalations containing menthol, camphor, and eucalyptus. Interestingly, modern research has shown that these substances do relieve symptoms to some extent, not by getting rid of blockage in the nose but by altering our perception of the quantity of air going through the nostrils. Although they are harmless, they are not particularly helpful and we don't recommend their use at all. Some people have tried acupuncture to relieve their symptoms. Again, such treatment does have some benefit, though quite how it works remains a mystery. It is possible that acupuncture diminishes the extent to which nasal blockage is appreciated by the brain rather than by having any effect on the nose itself. There are now many effective treatments for hay fever and which preparation is best depends in part on the severity of the symptoms.

Decongestants
Decongestants are used by millions of hay fever and rhinitis sufferers world-wide. Their use is rooted in antiquity. The type of

Table 3
Steps to successful treatment of allergic disease

Step 1	Find out if allergy is responsible for the symptoms (Chapters 5, 6, 7, and 8)
Step 2	Identify the causative allergens (Chapters 3 and 9)
Step 3	Take appropriate avoidance measures (Chapter 10)
Step 4	Match treatment to the severity of the symptoms (this chapter)
Step 5	Continue the treatment regularly for as long as is needed
Step 6	Keep fit and informed about the illness (this chapter)
Step 7	Aim to lead a full, active life; don't become a recluse or a food faddist

drug used in Sudafed, for example, is closely related to the active constituent of Ma Huang, a Chinese herbal medicine, which has been in use for more than 5000 years. All decongestants act by shrinking the blood vessels in the lining of the nose and work just as well for the nasal blockage occurring during the common cold as they do for that found in hay fever. Research in recent years has led to the introduction of compounds which are more powerful and longer lasting, such as oxymetazoline and xylometazoline, the active ingredients of Otrivine and Afrazine. The use of older compounds was often associated with progressive loss of effectiveness until, in the end, the very symptoms which they were designed to treat became more severe. It is doubtful whether this happens with modern formulations, though it is still best not to use these compounds for more than two or three weeks. These compounds can be taken in the form of tablets, though this greatly increases the possibility of side-effects. Decongestant tablets should not be taken by those suffering from high blood pressure or heart disease. Those taking decongestant tablets should check with their doctor that the tablets will not react with any other medication that may be taken.

Table 4
Hay fever treatment programme

Symptoms	Drugs	Problems
Mild	Decongestant sprays, drops and tablets Afrazine (oxymetazoline) Otrivine (xylometazoline) Sudafed (pseudoephedrine)	Avoid prolonged use; can increase nasal blockage.
	Antihistamines Traditional: Piriton (chlorpheniramine) Fabahistin (mebhydroline)	Some cause drowsiness and driving, operating machines, or drinking alcohol should be avoided.
	Modern: Zirtek (cetirizine) Triludan (terfenidine) Hismanal (astemizole)	The newer preparations do not cause drowsiness and will not interfere with everyday activities.
Medium	Antiallergics Opticrom Rhinacrom (sodium cromoglycate) Zaditen (ketotifen)	
	Topical steroids: sprays or drops Beconase (beclomethasone dipropionate) Betnesol (betamethasone sodium phosphate) Syntaris (flunisolide) Rhinocort (budesonide)	Occasional nose bleeds
Severe	Tablets or injections of steroids Prednesol (prednisolone) Kenalog (triamcinalone)	Steroid side-effects in large doses

Antihistamines
These tablets have been used for the past 50 years and act by opposing the action of histamine released from mast cells in the lining of the nose. They are particularly good for the treatment of

runny nose, sneezing, and running and itching of the eyes, but do not work so well against nasal blockage.

A great many antihistamines are now available, and this is an area in which advances are rapidly occurring. For many years the only available antihistamines, for example Piriton or Fabahistin, caused significant side-effects, especially drowsiness. This meant that people taking Piriton or Fabahistin could not drive, use machinery at work, or enjoy an alcoholic drink. Fortunately things have changed. Two new antihistamines, Triludan and Hismanal, are now available which do not cause any drowsiness for the vast majority of those that take them. Triludan can be bought over the counter without a prescription. It works very rapidly; symptoms should be relieved within an hour or two. Unfortunately, this drug does not stay in the body long and needs to be taken two or three times a day. Hismanal does not act as quickly, but has a more persistent effect and needs to be taken only once a day. Several new histamines are currently being developed and may provide even better relief of symptoms in the future. One compound that has recently become available is cetirizine—Zirtek. This is a potent antihistamine that acts quickly, only needs to be taken once daily, and does not cause drowsiness.

Unlike the combination tablets which contain more than one compound, some drugs on their own are thought to have several different actions. Zaditen, a compound widely used for the management of allergic disease throughout the world, is both a strong antihistamine and is thought to have additional anti-allergic activity. Although this drug is effective for hay fever, it has the disadvantage that it causes drowsiness.

Antiallergic drugs

A different way to tackle the problem of allergic rhinitis is to try to prevent allergen interacting with mast cells in the lining of the nose, thus stopping the release of histamine and other powerful chemicals. The first compound of this type to be made available for the treatment of allergic rhinitis was sodium cromoglycate, marketed under the name of Rynacrom. This compound is available in the form of drops, powder, or aerosol spray and is effective in the management of allergic rhinitis. Since it acts by preventing release of histamine and chemicals from mast cells it must be taken before contact with the allergen has occurred. For this reason patients with hay fever should start taking Rynacrom in the middle of May before grass pollen is present in the atmosphere in substantial amounts. If tree pollen is the cause of symptoms, treatment should be started in February. Rynacrom is not as effective in the management of perennial rhinitis due to allergy to domestic pets or house dust mite.

Sodium cromoglycate in the form of Opticrom is particularly effective for allergic conjunctivitis. Antihistamine drops should not be used since they themselves can cause the development of allergy in the eye. Steroid drops can cause side-effects if used long term, including an increase in the pressure in the eye (glaucoma), damage to the cornea, and even the development of cataracts. Unfortunately, the duration of action of Rynacrom and Opticrom is short, so they have to be taken 3–6 times a day. Another drawback is that Opticrom eyedrops contain a preservative called benzalkonium chloride which itself can cause an allergic-type reaction in some people. This additive can also react adversely with soft contact-lenses which should not be worn during treatment with Opticrom.

Topical corticosteroids

Of all the treatments available for allergic rhinitis the most effective are the corticosteroids. Unfortunately steroids have side-effects when taken in high dose for a long period of time. A great advance in use of corticosteroids for allergic disease was the development of compounds which could be inhaled in low dose into the nose, or lungs. These compounds are only active on the surface lining of the nose and lungs: they are rapidly broken down when absorbed into the body. This means that they exert their maximal activity where it is needed without being associated with side-effects in the rest of the body. There are three widely used preparations: Beconase, Syntaris, and Rhinocort. Although these compounds are very effective in the treatment of rhinitis and allergic conditions in general, their mode of action is not fully understood. Recent research supports the idea that steroids dampen allergic reactions by decreasing the number of mast cells and eosinophil cells in the lining of the nose. As for antiallergic drugs, it is best to start topical steroid sprays early in May for grass pollen allergy, or in February for tree pollen allergy. These topical steroids are useful in the treatment of perennial rhinitis and can prevent the regeneration of nasal polyps after operation. Side-effects, such as nose bleeds related to local irritation by the aerosol spray, occur in up to one in 20 sufferers using these sprays. Sprays which deliver the drug in liquid form rather than as a powder may have fewer side-effects. Betnesol nasal drops contain a higher-dose preparation of steroids and are useful in more severe cases.

Tablets and injections of steroids

To alleviate symptoms rapidly and effectively, for example before important social events or examinations, the best treatment is to take a regular daily tablet of a steroid called prednisolone, or an injection of another similar steroid, triamcinolone, in a long-lasting form marketed as Kenalog.

This type of treatment is also used by hay-fever sufferers for whom other treatments have proved ineffective. Their symptoms can be relieved throughout the pollen season by a single injection given when allergic symptoms appear.

Although these treatments are very effective, they cause suppression of the adrenal gland which normally produces steroids in the body. This will not give rise to side-effects if the course of prednisolone is short, for example one 5 milligram tablet daily for three or four weeks, or one injection of Kenalog in the summer-time for hay fever. However longer courses or more than one injection can cause the side effects outlined on page 176.

Nasal sprays and drops

If nasal sprays or drops are going to work they must get right into the nose. Although nose drops are frequently prescribed, they are often taken in an ineffective and incorrect way. Most people put drops into their nostrils with their head tipped back. The drops run down the floor of the nose and are soon tasted at the back of the mouth. In consequence the treatment is of no benefit. It is vital to get the drops right up into the nose and then for the fluid to spread throughout the nasal cavity. This will be helped by the flow of secretion in the nose which is normally driven backwards towards the throat by the little hair-like projections (cilia) on the surface of the lining of the nose. Nose drops should be taken as shown in Fig. 33. This position is known as the 'head down and forward' position and can either be achieved by kneeling on the floor with the top of the head resting on the floor, or by lying face down with the head hanging over the edge of the bed. The person's head must be kept in this position for about 30 seconds to allow the drops to get right up into the nose. By using this method it may be found that the drops that have been used for many years suddenly become effective.

Similarly as much as possible of the aerosol delivered from metered-dose canisters or from spray pumps must get deep into the nose. When the metered-dose canister or spray is activated, breathe in gently, closing off the opposite nostril with a finger. Two sprays into each nostril is the usual recommended dose. The nozzle of the spray must be directed straight back into the nostril to avoid deposition of the drug on to the nasal septum (which separates the left from the right nostril).

One of the problems about treating allergic rhinitis is that most sufferers wait until symptoms are bad before going to their doctor for treatment. By this time the nose is often runny, so that any drops are rapidly washed out by secretions, or the nostrils are too blocked for any aerosol or spray to enter the nose. Under these circumstances other treatment is needed first, to open the nostrils and reduce the amount of secretions. It is best to start with a rapidly acting antihistamine, such as one tablet of cetirizine (Zirtek) daily to

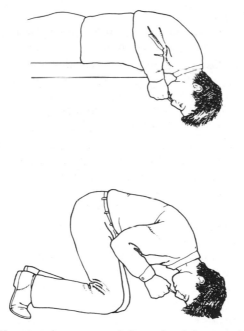

Fig. 33 How to take your nasal drops: head down and forwards position.

dry up the secretions, together with a nasal decongestant. Drops of Otrivine will rapidly cause the nasal lining to shrink, opening up the nose, so that within a day or two Beconase or Rynacrom can be taken.

Combination drugs

Combinations of antihistamines and decongestants for the treatment of allergic rhinitis, such as tablets of Hayphryn, Otrivine-antistin, and Antistin-privine, can be bought direct from the chemist. These are also available as nasal sprays, while Dimotane plus is available as a tablet. The antihistamines present in combination sprays may themselves cause allergy when regularly sprayed into the nostril. Similarly, combination tablets can be the cause of side-effects, particularly in people suffering from raised blood pressure or heart disease. These combination drugs are not recommended.

Hyposensitization

The type of treatment often called desensitization or 'allergy shots' is widely prescribed throughout the world, though its use in the United Kingdom is very limited. The aim of hyposensitization is to make people less allergic. This treatment can be very effective, but only a small proportion of people benefit. It works particularly well for only a few allergens, namely grass pollens, ragweed, and house dust mite. Results are better in hay fever and perennial rhinitis than in asthma. In bee- and wasp-sting allergy, hyposensitization is life-saving.

How hyposensitization works is not entirely clear. We do know that skin-test reactions, and the level of specific IgE antibodies in blood, decrease after successful treatment. In addition, IgG antibodies in blood increase. This is thought to represent the development of blocking antibodies, with which allergen reacts before it is able to combine with IgE antibody on the surface of mast cells. Hyposensitization may also work by increasing the production of protective antibodies of IgA class and by decreasing the number of mast cells in the lining of the nose.

Treatment involves the injection of purified allergen extracts below the skin. For seasonal allergens these injections need to be given before the season starts. The number of injections that need to be given differs according to the preparation used, but is usually between three and ten. The injection of allergen into someone who is allergic would be expected to cause reactions, and itchy lumps in the skin at the injection site commonly occur. Unfortunately, some patients died as a result of severe reactions to the injections, before strict guidelines for their administration were imposed. The Committee on the Safety of Medicines recommends that hyposensitization injections only be given where facilities for resuscitation are present, and where patients receiving the injections can be watched carefully for two hours following the injection. This has limited the administration of these injections to hospital clinics. In our opinion, hyposensitization is not as effective a treatment as the use of modern non-sedating antihistamines and topical corticosteroids, except in the case of allergy to insect stings. The number of individuals who are allergic to insect stings is small and there are only one or two centres in the country where such hyposensitization is administered.

Allergy to wasps and bees can be proved by detecting IgE antibody against the venom in blood, using the RAST test. Hyposensitization is recommended when RAST is positive and previous stings had led to severe reactions, for example the development of dizziness, widespread urticaria, or loss of consciousness. Treatment consists of seven or eight injections initially, followed by monthly injections for life. An alternative to hyposensitization is self-administration of adrenalin by injection immediately after a sting. Obviously the injection of adrenalin must be carried with the allergic person all the time, and these people must learn how to inject themselves, just as patients with diabetes learn to inject insulin.

Eczema and rashes

Atopic eczema

Many children suffer from dry skin. Soothing creams applied to the skin after bathing, or used instead of soap often help to relieve symptoms. Aqueous creams or emulsifying ointments are popular but many contain wool fat and related substances, including lanolin, which are known to cause sensitization, and also many preservatives to minimize bacterial growth. One of the most widely used and safest is simply called 'aqueous cream' which contains an emulsifying ointment with phenoxyethanol in freshly boiled and cooled purified water.

Modern ointments are occlusive and will prevent water getting on to the eczematous area. They most commonly consist of soft paraffin and have a mild anti-inflammatory effect themselves. Avoid ointments containing wool fat (lanolin) and alcohols. Some preparations combine both the qualities of creams and ointments in that they are greasier and more occlusive than creams but wash off readily with water; one such is Unguentum Merck which is very popular as an ointment. This material is often used as a base in which to put the active treatment for eczema.

Emollients sooth, smooth, and hydrate the skin, and are particularly helpful in dry eczematous disorders. It is often helpful to put emollients into the bath, and one commonly used emollient bath additive is made of oat protein (Aveno Regular), sometimes with some liquid paraffin added (Aveno Oilated). The use of perfumes, soaps, and bubble baths obviously should be avoided. Sunny holidays at the sea will often improve atopic eczema, as can a course of ultraviolet light. If these simple measures do not work, then it is worth considering a diet excluding dairy products (see Chapter 10).

Household pets commonly cause eczema in children, and may need to be removed. If symptoms persist, a steroid cream, of which a great many are available, may be necessary. Creams are easy to apply and less greasy than ointments, though the latter have advantages in that they can prevent cracking of dry skin. Lotions are particularly useful for application to areas of skin covered by hair. It is best to start with the weakest of the steroids, namely hydrocortisone cream now available direct from the chemist, for example Timocort. If the eczema does not clear within one or two weeks more potent compounds, for example clobetasone butyrate, known as Eumovate, should be tried. Even more potent steroid creams include Betnovate or Dermovate. Steroid creams combined with an antibiotic are necessary if infection has occurred in the eczematous areas: Betnovate N contains the antibiotic neomycin, and Betnovate C contains the antifungal agent, clioquinol. Prolonged use of potent steroids should be avoided, since thinning of the skin, stretch marks, red lines, acne, and additional hair growth can occur. Children suffer from disturbed sleep due to the irritation caused by eczema and benefit from a mild sedative antihistamine, such as Phenergan, given before going to bed.

Urticaria

Many cases of urticaria are related to reactions to foods, drugs, inhalant allergens, insect stings, and infections. Over-activity of the thyroid—thyrotoxicosis—and increased production of red cells, called polycythaemia, can also bring on attacks. If possible, the causative agent should be identified: this may be strawberries, mackerel, aspirin or other non-steroidal anti-inflammatory drugs, tartrazine, other colourings, and preservatives. Often it is not possible to identify such a precipitant and eliminate it, and additional treatment, with antihistamines, is needed. Non-sedative antihistamines, such as Triludan and Hismanal, are the drugs of choice. An alternative antihistamine that is often used is hydroxyzine (Atarax), though it has the drawbacks that it is sedative and can cause a dry mouth. A derivative of hydroxyzine, cetirizine (Zirtek), is now available and may prove very effective in the treatment of urticaria. Zirtek does not have any sedative effects, does not cause a dry mouth, and only needs to be taken once a day. It is potent and may influence other aspects of the allergic response. Unless the urticaria settles very rapidly, a diet low in azodyes, preservatives, salicylates, and colourings, and sometimes yeast as well should be tried. Painkilling tablets should be avoided. If the urticaria persists, a short course of treatment with prednisolone may be necessary— three or four tablets daily for a week or two. Although this dosage will temporarily suppress the adrenal glands' production of steroids, this will soon start again when the tablets are stopped.

In some people urticaria occurs around the lips, eyes, tongue, and larynx as a swelling rather than redness and itching. This is called angio-oedema and needs rapid and effective treatment. Many find that treatment with antihistamines is sufficient. If the attacks are severe, then treatment with tablets of steroids is vital, and in an emergency injections of adrenalin may be required. Depending upon the severity of the reaction half or one millilitre of adrenalin at a concentration of 1:1000 should be injected subcutaneously or even intramuscularly if the angio-oedema is particularly severe. Injections of adrenalin for self-administration are available in the form of Min-i-Jet in which the adrenalin is already in the syringe. Those carrying such medication have to learn to give themselves a subcutaneous or intramuscular injection. In a very small proportion of people who suffer from angio-oedema there is a deficiency of a protein in the blood, known as C_1-esterase inhibitor, which can be corrected by giving purified C_1-esterase inhibitor intravenously. Long-term preventative treatment involves the use of a drug called Danazol which, though effective, does have virilization effects in women, producing acne, hair growth, and a deepening of the voice. It is only sensible to take this drug in the case of frequent severe attacks of angio-oedema.

Asthma

Although asthma cannot be cured, the use of modern treatment enables most asthmatics to live a full and active life. No one likes taking treatment and, unfortunately, many with asthma wait until their symptoms are severe before taking medication. Although this is understandable, it is the wrong way to treat asthma. Prevention of symptoms in the first place is better than trying to treat them when they occur. The type of treatment, and how often it should be taken, depends on the severity of the asthma. Asthma ranges in severity from the occasional cough and wheeze on exercise to severe attacks requiring hospital admission, and can be divided into five stages.

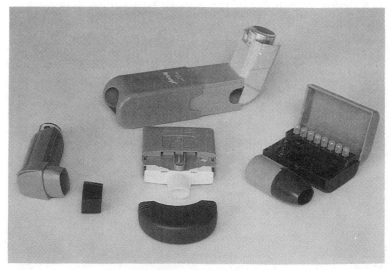

Fig. 34 Different types of inhaler.

The majority of people have mild asthma, stages 1 and 2, and this probably does not worsen throughout their life. In others asthma is severe from the very start, requiring frequent hospital admissions. As yet we do not know whether some patients whose asthma is mild initially gradually progress to severe attacks, and whether this progression can be prevented by modern treatment.

Stage 1

The great majority of asthmatics suffer only occasional attacks of wheezing, coughing, or shortness of breath brought on by a number of different circumstances, such as:

● exercise

● smoky atmospheres

● air pollution

● cold air or changes in temperature

● hair sprays, deodorants, perfumes

● paint fumes, detergents, disinfectants, bleaches

The reaction to environmental factors is a result of increased responsiveness of the lung airways, a feature known as bronchial hyper-responsiveness discussed in Chapter 2. Such symptoms will respond rapidly and readily to treatment with bronchodilators.

Bronchodilators These are the most common drugs used in the treatment of asthma. They can be inhaled either as an aerosol or as a powder or be taken in tablet form. The advantage of inhalers is that much lower doses of the drug are required since treatment goes direct to where it is needed. When the drug is taken as a tablet it is distributed to most parts of the body via the bloodstream, so that large doses must be taken to ensure that sufficient gets to the lungs, making it much more likely that side-effects will occur. Most bronchodilators are chemicals which relax the muscle around the airways of the lung and work in exactly the same way as adrenalin that naturally circulates in the blood. These compounds are described as beta-adrenoreceptor agonists. The importance of their action, at the specific receptors on smooth muscle, is well illustrated by the fact that drugs that block these receptors, known as beta-adrenoreceptor blocking agents, can precipitate severe attacks of asthma. A good example of such a drug is propranolol which is widely used throughout the world in the treatment of high blood pressure. Great care must be taken not to give this otherwise very useful compound to people who have asthma. Taking these beta-adrenoreceptor stimulants is associated with a number of side-effects, which if the compound is taken in aerosol or powder form are very minor. Many people notice a faint tremor of the hands initially, though this rapidly goes away. Similarly, palpitations occur in some people. Others develop headaches.

The most commonly used bronchodilators of this type are:

● salbutamol (Ventolin, Salbulin)

● terbutaline (Bricanyl)

● fenoterol (Berotec)

● reproterol (Bronchodil)

● rimiterol (Pulmadil)

The best way to take this type of therapy is from a metered-dose inhaler. This requires co-ordinating activation of the inhaler with breathing in, as shown in Fig. 35. Even when the metered-dose inhaler is taken correctly, only 15 per cent of the drug gets into the lungs—the rest is swallowed. This type of drug has no activity on the throat or in the stomach. Children under the age of five or six, and the elderly, or those with arthritis affecting the hands can find it very difficult to use metered-dose inhalers. Luckily there are plenty of alternatives, including rotahalers, turbohalers, and diskhalers in which the drug in powdered form is inhaled as the patient sucks on the mouthpiece. A whistle can be attached to some of these to encourage children to breath in deeply. Minor difficulties with co-ordination can be overcome by the use of a spacer device on a pressurized aerosol which makes it unnecessary to co-ordinate activation of the aerosol with inhalation. Interestingly, on a world-wide scale, bronchodilators are taken more frequently as tablets than as aerosols or powders, since many find the convenience of taking a tablet preferable to having to use an inhaler in public.

Great improvements have recently been made in the formulation of bronchodilator preparations. These rely on the slow release of the drug from a tablet or capsule. Such preparations, for example salbutamol controlled-release (Volmax), are likely to replace the more old-fashioned tablets of Ventolin or Bricanyl. Syrups are available for the treatment of infants.

In the past, the use of bronchodilator inhalers caused side-effects since the drugs they contained were not as specific as the modern ones and caused stimulation of the heart. This meant that their use needed to be carefully monitored. In contrast, the modern preparations described in this chapter are safe, non-addictive, and can be used frequently. However, if the bronchodilator aerosol is being

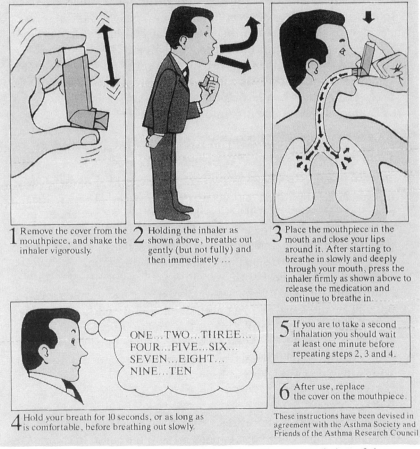

How to use your inhaler properly...

1 Remove the cover from the mouthpiece, and shake the inhaler vigorously.

2 Holding the inhaler as shown above, breathe out gently (but not fully) and then immediately ...

3 Place the mouthpiece in the mouth and close your lips around it. After starting to breathe in slowly and deeply through your mouth, press the inhaler firmly as shown above to release the medication and continue to breathe in.

ONE...TWO...THREE... FOUR...FIVE...SIX... SEVEN...EIGHT... NINE...TEN

5 If you are to take a second inhalation you should wait at least one minute before repeating steps 2, 3 and 4.

6 After use, replace the cover on the mouthpiece.

4 Hold your breath for 10 seconds, or as long as is comfortable, before breathing out slowly.

These instructions have been devised in agreement with the Asthma Society and Friends of the Asthma Research Council.

Fig. 35 How to use your inhaler (reproduced with permission of the Asthma Research Council).

used more than four or five times in the day, then additional therapy of the type described for more severe stages of asthma is required.

Xanthine bronchodilators   These drugs, namely theophylline and aminophylline, can only be taken as tablets. Their principle action is to cause relaxation of the muscle around the airways of the lung, but they also stimulate muscles in the heart, causing palpitations. Unfortunately, these tablets are associated with stomach upsets in as many as a quarter of those individuals who take the treatment, especially when they are taken in sufficient dose to exert maximal bronchodilator effect. More troublesome is the effect of these compounds on the brain, causing sleep disturbance which may lead to daytime tiredness. At very high doses, epileptic fits can occur. Because of this substantial list of side-effects these drugs are not frequently prescribed in the United Kingdom. They can, however, be of considerable help in individuals who suffer mostly from night-time attacks of asthma when the use of a slow-release preparation taken just at night can be helpful.

Examples of xanthine bronchodilators are:

● aminophylline (Phyllocontin)
● theophylline (Nuelin, Uniphyllin)

Anticholinergics   The nerves that control the calibre of the airways of the lung are called parasympathetic nerves. They release a chemical substance, called acetylcholine, which causes contraction of the muscle around the airways. This effect can be blocked by the use of an anticholinergic substance. The one that has been known since antiquity is atropine, but this has substantial side-effects, causing dryness of the mouth, palpitations, and severe psychological disturbance. In recent years an anticholinergic agent called ipratropium bromide (Atrovent) has been developed which can be inhaled into the lungs as an aerosol. Its great advantage over atropine is that it is very poorly absorbed and so side-effects are rare. This drug has proved useful in relieving narrow airways in patients not only with asthma but also with chronic bronchitis.

Stage 2 (seasonal asthma)

Some people only suffer attacks of asthma at certain times of the year, for example when pollen grains get into the lungs. In contrast to those who only suffer intermittent attacks which can be relieved by the occasional use of bronchodilators, people with seasonal asthma need to take therapy on a more regular basis during the summertime. If symptoms are mild, the regular use of a bronchodilator is all that is required. If this does not control symptoms, then additional treatment with an antiallergic preparation, such as sodium cromoglycate (Intal) or nedocromil sodium (Tilade), is useful. These drugs act by preventing attacks of asthma. It was thought that their major mode of action was to prevent the release of chemicals from mast cells triggered by contact with allergen. It is now thought that the action of these types of drugs is more complex and they probably interrupt the inflammatory process that results from contact with allergen. Whatever their mode of action, they need to be taken on a regular basis, whether symptoms are present or not. Intal can be taken either as a powder, using a spinhaler, or as a metered-dose aerosol, whereas Tilade is only available as a metered-dose inhaler. Intal has virtually no side-effects and Tilade is similar, though it is associated with a bitter taste in the mouth. At present much research is devoted to the production of a drug with similar activity to Intal and Tilade but which is active in tablet form. One such preparation, ketotifen (Zaditen), is available and is thought to have some of these properties.

Stage 3 (frequent symptoms all year round)

Many asthma sufferers experience symptoms throughout the year, often at night when they wake up coughing, wheezing, and short of breath. Like those with seasonal symptoms, such asthmatics need regular preventative therapy. It is unwise to rely solely on bronchodilators under these circumstances, and it is better to try to prevent symptoms. Drugs such as Intal, Tilade, and Zaditen can be useful, but an alternative, particularly in severe cases, is the use of inhaled corticosteroids. Great advances have taken place in the development of these drugs in that they are very active when inhaled into the lungs and the small amount that is absorbed into the bloodstream is rapidly broken down so that side-effects are minimized.

Most people are upset at the prospect of taking regular steroids, even in inhaled form, but when taken in this way they are remarkably free of side-effects. The only real problems arise in the throat and larynx where they can cause the development of a fungal infection called thrush, and a hoarse voice. Fortunately, these effects only occur in about one in 20 people who regularly use inhaled steroids. The chances of developing such side-effects can be reduced by using the inhalers before meals or by rinsing the mouth out after inhalation. The use of the inhaled corticosteroids is very effective in keeping asthma of this type under control. If occasional attacks of wheezing, coughing, or shortness of breath still occur, for example after exercise, the addition of a bronchodilator causes rapid relief.

The mode of action of corticosteroids is complex and not fully understood. Almost certainly their effectiveness is due to the fact that they reduce the inflammation that occurs in the airways of the lung in this type of asthma.

Stage 4 (disabling symptoms)

In a small proportion of patients with asthma, symptoms continue in spite of treatment with regular preventative therapy in the form of Intal or inhaled steroids, and more treatment is required. Such asthma sufferers experience shortness of breath which interferes with their daily life, for example they become too breathless to carry shopping and find climbing stairs difficult. Sporting activities are out of the question. Vigorous treatment is required, usually with an increased dose of inhaled corticosteroids in the form of beclomethasone dipropionate (Becloforte) or budesonide (Pulmicort). In the case of Becloforte, which is the same steroid as is found in Becotide, the dose is increased fivefold. These compounds have proved very effective in helping more severe asthmatics, but even their use may not be sufficient and such asthmatics may require tablets of corticosteroids.

Tablets of corticosteroids are associated with many unwanted side-effects, though whether side-effects occur or not depends very much on the number of tablets that are taken on a daily basis. If the dose can be kept below two tablets (10 milligrams), side-effects are minimal and usually include an increase in appetite with associated weight gain and, in some patients, the development of acne. Stomach ache may also occasionally occur. The most widely used oral corticosteroid is prednisolone. Higher doses of this drug taken on a regular daily basis can, over the years, cause substantial side-effects, including the development of facial hair, easy bruising, thinning of the bones, stomach ulcers, and, in some people, the development of diabetes. Obviously, their use in high dose should be avoided if at all possible, although unfortunately sometimes asthma can be so severe that there is no alternative.

Many patients with regular disabling asthma find benefit from the use of bronchodilators, such as salbutamol and terbutaline in high dose administered using a nebulizer. This enables the asthmatic to undergo treatment with very much larger doses of these drugs than can be obtained from the use of metered-dose inhalers, rotahalers, or diskhalers. Nowadays nebulizers are portable, and can be taken on holiday, or used in the car, where the power comes from the car battery via the socket for the cigarette lighter. It is vital that asthmatics do not rely upon the use of nebulized bronchodilators alone when their asthma worsens but that they seek advice from their doctor. This is because the use of high-dose bronchodilators via a nebulizer only produces transient relief, and more fundamental therapy, with high doses of corticosteroids, may be necessary to stop asthma getting progressively worse.

Stage 5 (severe asthmatic attacks)

In spite of all the treatment that is currently available for asthma, occasionally the disease gets out of hand and medical help is essential. Under these circumstances all the treatment described so far will have proved ineffective, and hospital treatment is necessary. In hospital, regular nebulized bronchodilator treatment together with steroids given directly into the bloodstream by injection is necessary. In addition, oxygen is needed and, in extreme circumstances, it may become necessary to assist breathing with a mechanical ventilator. If the asthma is not responding to the usual medications, it is vital to call the doctor who may well recommend admission to hospital. Do not delay—remember, asthma can prove fatal.

Asthma support groups

Asthma, particularly if it is severe, can be a frightening illness and sufferers need to know as much as they can about the disease and its treatment. There is a network of asthma support groups throughout the country and details about them can be obtained from the Asthma Society and Friends of the Asthma Research Council, 300 Upper Street, London N1 2XX. Whatever the stage of asthma, it is essential to take sufficient therapy to lead an active normal life.

Many asthmatics are helped by exercise programmes, which help to instil a sense of confidence and improve fitness. Whatever the limitations imposed by asthma they can be improved by regular exercise. The combination of the right drug treatment, the right attitude, and regular exercise should mean that most asthmatics can look forward to a normal life.

APPENDIX A

IMMUNOLOGY — ANCIENT AND MODERN

The beginnings of immunology

Immunology really began at the end of the nineteenth century and is a relatively young branch of medical science, certainly when compared with anatomy. It is the study of immune responses, and the word itself is derived from the latin '*immunia*' meaning 'exempt from charges (taxes)', hence its use as a term denoting exemption or resistance to infectious diseases. Indeed, it was bacteriologists, or at least those interested in infectious diseases, who made the important early observations. The fact that some people are resistant to a second 'attack' of a particular disease has been known for centuries. In ancient China, protection against smallpox, known as variola, was sought by infecting people with blister fluid from patients suffering from a mild form of the disease. This process of 'variolation' was introduced into England from Turkey in 1721 by Mary Wortley Montagu, with varying results since, unfortunately, severe disease and death often resulted. It was the English physician Edward Jenner (1749–1823) who revolutionized the treatment of smallpox. He observed that people previously infected with the milder cowpox were protected against subsequent smallpox. He tested this in a courageous, but perhaps somewhat unethical, study by innoculating an eight-year-old boy, James Phipps, with fluid from a cowpox blister. The effectiveness of this process, called 'vaccination' from the Latin '*vacca*' — 'a cow', was immediately striking, although scorned at the time by the medical establishment. Nevertheless, he had discovered what must be considered an almost miraculous preventative therapy for a disease which at that time afflicted 60 per cent of the population of England and which now, thanks to this general practitioner from Gloucester, has been eradicated throughout the world.

The great French scientist Louis Pasteur (1822–95) extended these observations, realizing the possibility of protection against infection by 'vaccination' which, as with the use of the less dangerous cowpox virus, involved injections of less harmful and less virulent micro-organisms. By chance he discovered that 'dangerous' bacteria left to grow in test-tubes over a holiday had become much less infective and harmful—a procedure he used to produce vaccines against many bacteria causing important diseases at that time, such as anthrax and rabies. It was obvious that vaccination worked, the question was how? In the late nineteenth century it was appreciated that both cells and soluble non-cellular, or humoral, materials played very important roles in the defence systems of the body: these activities were increased in patients recovering from an infective illness and in people who had been vaccinated.

In 1882 the Russian zoologist Elie Metchnikoff introduced a thorn into the side of a transparent starfish larva and found that it was soon surrounded by motile cells. He went on to show that these white cells in the blood (leucocytes) could engulf micro-organisms such as bacteria, a process he described as phagocytosis. This is illustrated in Fig. 36, and is a process which could be performed by two different types of cells in the blood: one a cell in which the nucleus was characteristically lobulated into various shapes—a neutrophil granulocyte—and the other a larger cell with a more regular nucleus—the macrophage. However, a few years later von Behring and Kitasato, working in Berlin, showed that immunity could be transferred from a patient recovering from diphtheria or tetanus to another individual by injecting blood serum, that is blood without any of its cells. Similarly, the serum from animals injected with diphtheria or tetanus toxin could neutralize the effects of these agents producing disease in man. So began the era of serum therapy, and Institutes in Marburg, Paris, and London began the preparation of large quantities of 'antiserum' in horses for subsequent use in man. What was this component present in the 'antiserum'? Initially it was described in terms of the function it appeared to possess; for example agglutination of foreign matter, in which case it was called an agglutinin, or sensitizing an individual, when it was called a 'sensitizer'. Finally, all these functions were encompassed in the term 'antibody'. This is a body protein which can react with foreign materials, usually proteins—which are now called antigens.

Considerable argument raged at this time as to which was the most important factor in the protection of man from invading microbes, the 'cellular immunity' theory of Metchnikoff or the

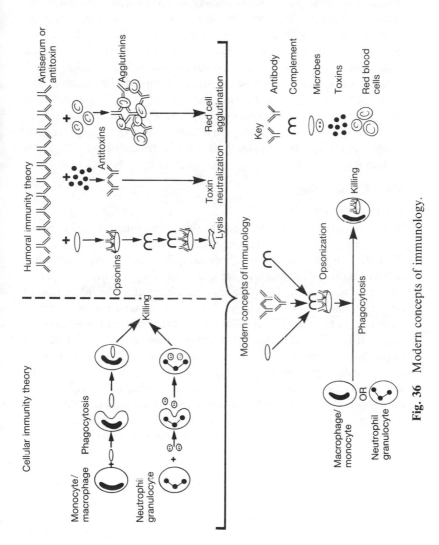

Fig. 36   Modern concepts of immunology.

'humoral' theory of the more modern investigators. A further complication arose in 1900 when a young Belgian, Jules Bordet, described an additional factor present in serum which helped in the breakup, or lysis, of bacteria. Its activity, however, was destroyed by heating the serum to 56 °C. This factor which he called 'alexine' became known subsequently as complement. It was left to the Englishmen Wright and Douglas working at St Mary's Hospital in London in 1903 to propose a bridge between these conflicting theories, suggesting that the antibodies present in the serum facilitated the phagocytosis of microbes by leucocytes. By this time, over 80 years ago, the basis of our current understanding of protection against microbes and toxins had been formulated and 'modern immunology' began.

Modern immunology

The immune response

It is now known that this remarkable mechanism present in the human body has several defining characteristics.

First, it is extraordinarily specific. That is, it can distinguish between minute differences in the make-up of a particular allergen or antigen. The individual may develop hay fever after the inhalation of one particular grass pollen but not after inhalation of another; asthma may be caused by inhalation of material from the house dust mite but not by inhalation of material from its close relative, the grain mite. An industrial worker exposed to chemicals may develop disabling asthma following the inhalation of the vapour of toluene di-isocyanate but not after inhalation of the closely related hexamethylene di-isocyanate. A baby can develop eczema from drinking cow's but not human milk. This phenomenon of specificity is absolutely fundamental to our understanding of allergy as a whole and is discussed throughout this book.

Secondly, the body acquires a memory of the first contact with the allergen so that subsequent responses are quicker and, indeed, amplified. This is well recognized in practice, for example newcomers to a particular environment may not develop allergic disease for one or two years, but then symptoms become progressively worse. It usually takes several bee stings, perhaps as many as 20 or 30, before severe reactions occur.

Thirdly, the allergic response is controlled so that it encompasses only foreign and not 'self' proteins. This is a critical part of the response, namely the maintenance of tolerance to one's own body, a phenomenon known as self-tolerance. In fact we now know that there are antibodies present in the blood directed against constituents of one's own body. Nearly always their production remains limited but sometimes control is lost and disease results, for example in some cases of over- and then underactivity of the thyroid gland, known as thyrotoxicosis and myxoedema respectively. Very occasionally this production of antibodies against tissues in the body becomes widespread, leading to a severe disease called systemic lupus erythmatosis (SLE) which affects many different organs of the body. Fortunately, these so-called 'autoimmune' diseases are rare. However, it is possible that more diseases than we currently think are caused by this autoimmunity. For example, no external agent, i.e. allergen, can be discovered which might cause hay-fever-like symptoms (rhinitis), asthma, or eczema in many patients. These diseases are often called 'intrinsic' rhinitis or 'intrinsic' asthma, implying that the cause is within the body itself. Indeed this may well be the case, even though at the present time no antibodies directed against the body's own cells (autoantibodies) in the nose or lung have been identified in these diseases.

The organs of the immune response

The organs of the immune response are situated throughout the body. They are generally referred to as lymphoid organs because they are concerned with the growth, development, and deployment of lymphocytes, the white cells that are critical to the functioning of the immune system. Lymphoid organs include the bone marrow, the thymus, the lymph nodes, and the spleen, as well as the tonsils, the appendix, and clumps of lymphoid tissue in the small intestine known as Peyer's patches.

Cells destined to become lymphocytes start their life in the yolk sac of the developing embryo in the womb, dividing and then migrating to the fetal liver. From there some of these cells travel to the soft tissues inside long bones, the bone marrow. Other cells migrate to the thymus, a multilobed organ that lies high behind the breastbone. There they multiply and mature into cells capable of producing an immune response; that is they become immuno-competent. Cells that develop in the thymus are called T-lymphocytes whereas those that mature either in the bone marrow or in lymphoid organs other than the thymus are called B-lymphocytes.

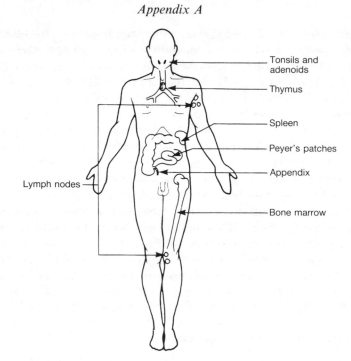

Fig. 37 Organs of immune response.

Lymph nodes are small, bean-shaped structures that are spread throughout the body. Strings of such nodes are found in the neck, armpits, abdomen, and groin. Each lymph node has specialized compartments, some housing B-lymphocytes, others T-lymphocytes, while other areas of the lymph nodes contain other immune cells, in particular the macrophages. Additionally, lymph nodes have net-like tissues which can 'trap' antigens. Thus, the lymph nodes and the other organs of the immune system contain all the necessary equipment to mount an immune response to foreign matter. Lymph nodes are linked by a network of lymphatic vessels similar to the blood vessels. These channels, at first minute, link like streams into ever enlarging vessels, eventually emptying into the blood at the base of the neck. Inside the vessel is lymph, a clear fluid that bathes the body's tissues. Lymph contains many cells, most of them lymphocytes. As lymph drains through lymph nodes, antigens are filtered out and more lymphocytes are picked up, eventually being carried into the bloodstream and hence around the body. In the bloodstream, T-lymphocytes outnumber B-lymphocytes by about five to one and T-lymphocytes move around the body much more than B-lymphocytes. This distribution of lymphocytes, and indeed macrophages, throughout the body means that they are ready and waiting to mount an immune response against foreign matter wherever it is required, and by returning to the organs of the immune response can recruit other lymphocytes and macrophages to help in the body's defence against invaders. We are all aware of this response when the lymph nodes in our neck enlarge, often becoming tender, in response to bacteria causing a sore throat. The importance of these lymphoid organs in resisting disease can be appreciated in individuals who have had to have their spleen removed following abdominal injury, since they become more susceptible to severe infections. The spleen, in fact, contains enormous numbers of lymphocytes, many of them in passage between the bloodstream and lymphatic system, and it can and does enlarge, particularly in response to parasites, such as those causing malaria, which circulate in the blood.

The cells of the immune response

Lymphocytes   The lymphocytes are the major cells involved in the immune system. They are present in very large numbers in the human body, in excess of a million million cells. The first and cardinal feature of both B- and T-lymphocytes is that they are capable of reacting with a particular molecular shape. Each lymphocyte will react with only one particular shape, just like a lock will accept only one key. It seems ai most incredible that there is present on the surface of each one of these cells a particular-shaped lock (receptor) awaiting the arrival of a particular molecule (antigen) of identical but inverse shape. The lymphocytes are capable of responding to millions of foreign antigens but not to self-proteins. This concept stemmed from the ingenious theory expounded in 1959 by McFarland Burnett, designated by him 'the clonal selection theory of acquired immunity'. He suggested that the ability to recognize a particular antigen was transmitted through many generations of lymphoid cells, and that each clone of those cells would thus produce only one specific type of antibody. The reason why a cell did not react with self-protein was due, he suggested, to the fact that such lymphocytes were eliminated or destroyed as forbidden clones during fetal development.

Antibodies are produced by one particular type of lymphocyte, the B-lymphocyte, and each of these cells is programmed to make only one specific antibody. When the B-lymphocyte encounters its triggering antigen it divides to give rise to many cells which mature into what are called 'plasma cells'. These cells are essentially a factory for producing antibody, and are all descended from one particular B-lymphocyte and therefore are all members of the same family or clone. Each plasma cell can produce up to 2000 antibody molecules per hour during their brief life-span of a few days.

T-lymphocytes do not secrete antibody, though their help is essential for antibody production. There are different types of T-lymphocytes; some, designated 'helper cells', help B-lymphocytes to transform into plasma cells and secrete antibody, while others have

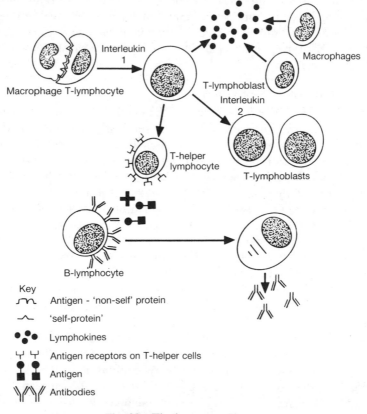

Key

Antigen - 'non-self' protein

'self-protein'

Lymphokines

Antigen receptors on T-helper cells

Antigen

Antibodies

Fig. 38 The immune response

the reverse function and are known as 'suppressor cells'. Such control mechanisms for B-lymphocyte proliferation and development into plasma cells, and antibody production, are essential for appropriate regulation of the response to foreign antigens. T-lymphocytes are essential in the first step of the immune response, when they interact with macrophages which have picked up antigen. They secrete a protein, interleukin 2, which causes them to proliferate and become active, secreting diverse and potent chemicals called 'lymphokines', which can attract and activate other cells as well as having direct functions in the development of inflammation. As already mentioned, all lymphocytes carry on their surface receptors that will recognize one particular antigen. In the case of the B-lymphocyte this receptor is a sample of the antibody that that particular B-lymphocyte is prepared to manufacture. In the case of T-lymphocytes this is a surface receptor which will lock directly with antigen. Further, it can interlock with and destroy cells that carry the corresponding antigen; these may be bacteria, or transplanted tissue cells, or body cells that have become infected by viruses. Indeed, such a mechanism may be very important in removing cells which have become cancerous.

Some lymphocytes have a natural killer function which, as their name suggests, indicates that they can attack and destroy other cells without prior stimulation by a specific antigen. Most normal cells are resistant to natural killer cell activity. However, most tumour cells, as well as normal cells infected with a virus, are susceptible. Thus, these natural killer cells may play a key role in immune surveillance against cancer, hunting down any cell that develops abnormal changes.

Macrophages and monocytes   These cells are large and can act as scavengers or phagocytes. They can engulf and digest marauding micro-organisms and other particulate antigens. Monocytes circulate in the blood while macrophages, in a variety of guises, are found throughout body tissues. Macrophages play a crucial role in initiating the immune response by presenting antigen to T-lymphocytes in a special way that allows the T-lymphocyte to recognize it as foreign. In fact, the T-lymphocyte recognizes a complex structure on the surface of the macrophage which is made up of a mixture of the cell's own protein and that of the foreign antigen. To attract further T-lymphocytes, the macrophage secretes a protein, interleukin 1, as shown in Fig. 38. In addition to this vital role in the immune response, both macrophages and monocytes secrete an amazing array of powerful chemical substances which are called 'monokines'. They are comparable to the lymphokines secreted by lymphocytes and, like lymphokines, they help not only in the attraction and activation of other inflammatory cells but also play vital roles in the development of the healing process in the body.

Granulocytes   These cells contain granules filled with potent chemicals that enable them to digest micro-organisms: however, these chemicals also contribute to the development of inflammation in tissues, and have a very important role in the development of reactions that cause allergic diseases, such as asthma and hay fever. The neutrophil polymorphonuclear granulocyte is so called because its nucleus has many shapes. It is a cell which responds dramatically to bacterial infection, increasing several-fold in number in blood and forming the major cell type in pus coming from infected wounds or boils. The eosinophilic granulocyte has a characteristic bi-lobed nucleus, and granules that stain a characteristic reddish-yellow with the dye eosin. This cell is a hallmark of allergic disease and of invasion of the body by parasitic worms. The basophil granulocyte contains granules which stain blue with basic dyes. It has many features in common with the tissue mast cell, though its origin is thought to be different. Whereas the basophil is thought to be derived from cells in bone marrow it now seems that mast cells may differentiate from precursor cells present in tissues which are more closely related to the lymphocyte series.

Antibodies and complement

It is now known that there are at least five classes of antibodies, which are usually called 'immunoglobulins', designated Igs. The five classes are: IgG, IgM, IgA, IgE, and IgD. Much is now known about the exact make-up of antibodies, following the pioneering work of Porter in 1959. He showed that antibodies could be split by the plant enzyme papain into three major fragments. Two of these fragments, known as Fab (fragment antigen binding), were identical and the third, known as Fc (fragment crystalline), was different. It was the Fab fragments that were able to combine with a foreign antigen, whereas the Fc fragment was responsible for certain properties of the immunoglobulin, such as binding of other cells. Each antibody

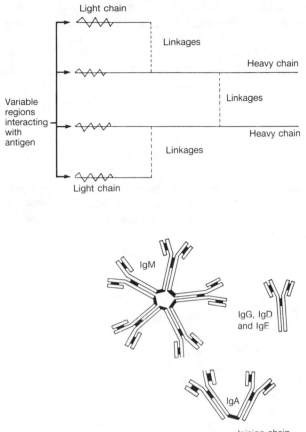

Fig. 39 Antibody (immunoglobulin) structure

molecule is now known to consist of two distinct types of protein chains, as shown in Fig. 39. It is the characteristics of the heavy chains that place each of the molecules into distinct immunoglobulin classes. In some parts of the light and heavy protein chains the constituent amino acids vary greatly and form the antigen-combining site of the antibody, an area usually composed of up to 17 amino acids.

Each antibody plays a different role in the immune mechanism. IgG is the antibody present in the largest amounts in the human body and is the major immunoglobulin in blood. It can also enter tissue spaces and works efficiently to coat bacteria for example, speeding their uptake by phagocytic cells. IgM is a cluster of five antibodies and, because of its large size, remains in the bloodstream where, in combination with complement in particular, it can directly kill bacteria. IgA is concentrated in body fluids, such as tears and saliva, as well as in the secretions of the respiratory and gastrointestinal tracts. It probably has a role in interacting with micro-organisms to prevent them from entering the body. It is present in secretions as two molecules joined together by a connecting piece. Little is known about the precise activity of IgD, although it appears to be inserted into the membrane of cells where it may have a role in regulating their activity. IgE has specially developed heavy chains which allow it to fix strongly to the surface of basophil and mast cells. It is this antibody which is central to the development of allergic diseases.

Although antibodies cannot penetrate living cells, they are present throughout the body and its surface. Whereas some antibodies will react with toxic chemicals to inactivate them, others coat the surface of micro-organisms to aid their phagocytosis. Some will prevent viruses from entering cells, and others are able to kill bacteria directly. This latter function, carried out by IgG and IgM in particular, relies upon a further interaction with a series of 20 or so proteins that normally circulate in blood in an inactive form. These comprise the complement system. When the first of the complement substances is triggered by interaction of antibody (IgG or IgM) with an antigen, forming an antigen–antibody complex, a cascade of sequences is set up. Each component of complement is activated as it in turn acts upon the next in a precise sequence of carefully regulated steps. This complement cascade can lead eventually to the release of proteins which break down cell membranes, for example those of bacteria, causing their death. Another part of the complement cascade can directly activate mast cells, causing allergic-like reactions which have not been induced by the interaction of allergen with IgE antibody. Indeed, the components of the complement cascade which have this function are known as anaphylatoxins, and it is release of these components, for example, that causes asthma and low blood pressure in some individuals who react to intravenous injections of X-ray-opaque contrast media used to study organs such as the kidneys.

The induction of an immune response

On a world-wide basis, infections caused by bacteria, viruses, parasites, and fungi remain the most common causes of human disease. Infections can range from relatively mild respiratory illnesses such as the common cold, to debilitating conditions such as hepatitis, and life-threatening diseases such as meningitis and malaria. The body has devised a series of highly intricate reactions to prevent such micro-organisms from entering the body, multiplying, and finally killing their host. Such microbes as manage to cross the skin or the mucous linings of the respiratory or intestinal tract are ingested by the macrophage. This cell will display the characteristic markers of the foreign microbe on its surface next to a part of its own protein surface structure. This combined shape, part self and part non-self, is necessary for recognition and then interaction with the T-lymphocyte. Once triggered, the T-lymphocyte enlarges and releases chemicals known as interleukins which carry messages to other T-lymphocytes, inducing them to enlarge and become active. Some of these cells are then capable of killing the invading micro-organisms. Others synthesize lymphokines which will draw more macrophages (in particular), as well as granulocytes, to the site of infection. At the same time, appreciation of the foreign antigen by antibody expressed on the surface of a particular B-lymphocyte leads to its stimulation and the development of a clone of B-lymphocytes producing antibody against the invading organism. This transformation of the B-lymphocyte into the antibody-producing plasma cell is helped by T-helper lymphocytes. This response leads to the death of invading microbes and, under the influence of suppressor factors produced by yet another group of T-lymphocytes, suppressor T-lymphocytes, antibody production wanes.

Appendix B

BLOOD TESTS FOR ALLERGY

PRIST testing

In this test a paper disc is first coated with antibodies made in an animal, usually a rabbit, against human IgE. This is called anti-human IgE. Blood is then added and any IgE antibody in the blood will stick to the anti-human IgE on the disc. After washing, a solution containing more anti-human IgE is added, but this time the anti-human IgE has been labelled with a radioactive marker. This will stick to the patient's IgE, making a three layer 'sandwich' on the disc. The amount of IgE originally in the patient's blood can be estimated from the amount of radioactivity bound to the disc. The principle of the test is shown in Fig. 40.

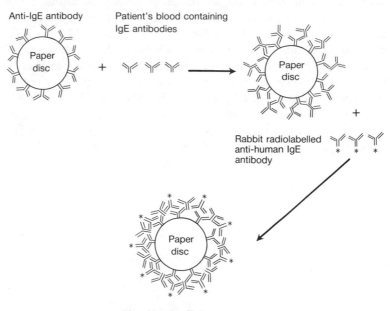

Fig. 40 Radioimmunoassay.

RAST testing

In this test very pure extracts of allergen (for example grass pollen or house dust mite), rather than the anti-human IgE antibody as in the PRIST testing, are stuck to the surface of the paper disc. These discs are covered with the patient's blood. If any antibodies (specific IgE) to the allergens are present, they will stick to the allergen on the disc. Once any excess serum has been washed away a further solution is added, again containing anti-human IgE against the antibodies which are already attached to the disc. These radioactively labelled antibodies attach to the specific IgE already bound to the allergen on the disc and cannot be washed away. Once any extra solution has been washed away the amount of specific IgE in the patient's blood can be estimated from the amount of radioactivity which remains stuck to the disc.

MAST testing

The 'chemiluminescent assay' (CLA) measures both total and specific IgE in human blood. In the MAST (multiple allergosorbent test) system, anti-human IgE attaches to the patient's antibodies as in the RAST test. In this case, however, the anti-human IgE is labelled not with radioactivity but with an enzyme which reacts with other chemicals causing luminescence. This can be photographed using high-speed film. Instead of the test allergens being on paper discs, as for RAST, in the MAST system they are attached to a thread contained in a special test chamber.

INDEX